Integrated Aquifer Characterization and Modeling for Energy Sustainability

The greatest challenge facing humanity today is the transition to a more sustainable energy infrastructure while reducing greenhouse gas emissions. Meeting this challenge will require a diversified array of solutions spanning across multiple industries. One of the solutions rising to the fore is the potential to rapidly build out carbon sequestration, which involves the removal of CO_2 from the atmosphere and its storage in the subsurface. *Integrated Aquifer Characterization and Modeling for Energy Sustainability: Key Lessons from the Petroleum Industry* provides a comprehensive and practical technical guide into the potential that aquifers hold as sites for carbon and energy storage.

Aquifers occupy a significant part of the Earth's available volume in the subsurface and thus hold immense potential as sites for carbon storage. Many aquifers have been studied extensively as part of oil and gas energy development projects and, as such, they represent an opportunity to sequester carbon within existing areas of infrastructure that have already been impacted by, and integrated into, an inherited energy framework. Moreover, future efforts to reconfigure the landscape of our national and global energy systems can extract valuable lessons from this existing trove of data and expertise.

From a multidisciplinary perspective, this book provides a valuable and up-to-date overview of how we can draw on the wealth of existing technologies and data deployed by the petroleum industry in the transition to a more sustainable future. *Integrated Aquifer Characterization and Modeling for Energy Sustainability* will be of value to academic, professional and business audiences who wish to evaluate the potential underground storage of carbon and/or energy, and for policy makers in developing the right policy tools to further the goals of a sustainable energy transition.

Integrated Aquifer Characterization and Modeling for Energy Sustainability

Key Lessons from the Petroleum Industry

M.R. Fassihi and J.P. Blangy

CRC Press
Taylor & Francis Group
Boca Raton London New York

CRC Press is an imprint of the
Taylor & Francis Group, an **informa** business

A BALKEMA BOOK

Cover image: Kelly Wrobel

First published 2023
by CRC Press/Balkema
Schipholweg 107C, 2316 XC Leiden, The Netherlands
e-mail: enquiries@taylorandfrancis.com
www.routledge.com – www.taylorandfrancis.com

CRC Press/Balkema is an imprint of the Taylor & Francis Group, an informa business

© 2023 M.R. Fassihi & J.P. Blangy

The right of M.R. Fassihi & J.P. Blangy to be identified as authors of this work has been asserted in accordance with sections 77 and 78 of the Copyright, Designs and Patents Act 1988.

All rights reserved. No part of this book may be reprinted or reproduced or utilised in any form or by any electronic, mechanical, or other means, now known or hereafter invented, including photocopying and recording, or in any information storage or retrieval system, without permission in writing from the publishers.

Although all care is taken to ensure integrity and the quality of this publication and the information herein, no responsibility is assumed by the publishers nor the author for any damage to the property or persons as a result of operation or use of this publication and/or the information contained herein.

ISBN: 9781032224954 (hbk)
ISBN: 9781032224992 (pbk)
ISBN: 9781003272809 (ebk)

DOI: 10.1201/9781003272809

Typeset in Times New Roman
by Newgen Publishing UK

To
Theresa
and
Pamela

Contents

Preface   xiii
Acknowledgments   xv
Author biographies   xvii

1 Introduction   1

2 Aquifer presence and identification   5

Introduction   5
2.1 *Definitions*   6
2.2 *Data availability, cost and schedule*   7
    2.2.1   Fields in the late exploration or appraisal phase   8
    2.2.2   Fields under development and/or production   8
2.3 *Detection methods: direct detection*   10
    2.3.1   Surface sampling and mapping   10
    2.3.2   In situ sampling or well-based methods   10
       2.3.2.1   Well logs   10
       2.3.2.2   Pressure and temperature   10
       2.3.2.3   Rock and fluid samples   11
2.4 *Detection methods: remote sensing*   14
    2.4.1   Potential fields   14
    2.4.2   Seismic   15
       2.4.2.1   Seismic data characteristics   15
       2.4.2.2   Seismic data type, availability and quality considerations   16
       2.4.2.3   Seismic resolution and seismic detection from reflections   19
       2.4.2.4   Seismic observations   20
2.5 *Examples of aquifer detection and identification*   21
    2.5.1   Atlantic margin, offshore United States   21
       2.5.1.1   Well data acquired in the 1970s and the 2010s   21

		2.5.1.2	New well and seismic data acquired by IODP-313 in the 2010s	24
		2.5.1.3	Modern potential fields data acquired in the late 2010s	25
	2.5.2	A seismic example: offshore Angola		27
		2.5.2.1	Descriptive seismic attributes	27
		2.5.2.2	Using seismic to identify and map the aquifer	28
	2.5.3	Detection of a hydrodynamic system		29
		2.5.3.1	The Cretaceous Nahr Umr Lower Sands, offshore Qatar	29
2.6	Aquifer extent summary and conclusions			31

3 Aquifer description and characterization — 37

Introduction — 37

3.1	Aquifer characterization using geological information		37
	3.1.1	Regional geology: a play-based approach	37
		3.1.1.1 Play-based concepts	37
	3.1.2	Depositional systems and stratigraphy	40
	3.1.3	Structural model and faulting	43
		3.1.3.1 Large-scale structural deformation	46
		3.1.3.2 Local scale structural features and fault traps	47
	3.1.4	Reservoir quality assessment	48
		3.1.4.1 Thin sections	49
		3.1.4.2 Core analyses	49
	3.1.5	Production geology	50
3.2	Aquifer characterization using geophysical information		51
	3.2.1	Internal geometry of the aquifer: stratigraphy	52
	3.2.2	Internal geometry of the aquifer: faulting	53
	3.2.3	Gas–water contact identification	55
	3.2.4	Oil–water contact identification	55
	3.2.5	Rock-physics-based seismic inversion for facies classification	56
	3.2.6	Production geophysics and surveillance	60
3.3	Aquifer characterization using petrophysical information		63
	3.3.1	Petrophysical interpretation	64
	3.3.2	Transition zones	68
		3.3.2.1 Capillary pressure	68
		3.3.2.2 FWL versus OWC	68
	3.3.3	Paleo zone	70
	3.3.4	Tar mat	77
3.4	Anomalous water		79
	3.4.1	Perched water	80

	3.4.2	Quick checklist for perched water identification	82
	3.4.3	Hydrodynamically tilted contacts	83
3.5	Summary		88

4 Static models for reservoirs and their aquifers 93

Introduction 93

4.1 Basin models 93
    4.1.1 Definition and functionality 93
    4.1.2 Applying physics: solving for pressure and temperature 94
    4.1.3 Current challenges and limitations in basin modeling 99
        4.1.3.1 Coupled structural evolution/basin history 99
        4.1.3.2 Multi-phase compositional fluid flow 99
        4.1.3.3 Diagenesis 99

4.2 Reservoir quality models 100
    4.2.1 Porosity models 101
    4.2.2 Permeability from porosity 104

4.3 Geomodeling: field scale integration 105
    4.3.1 A typical geomodeling workflow 105
        4.3.1.1 Phase 1: building the structural and stratigraphic framework 106
        4.3.1.2 Phase 2: assigning layer properties 110
    4.3.2 Aquifer-specific geomodeling 112
    4.3.3 Preliminary volume calculations and volumetric uncertainty assessment 114
        4.3.3.1 Basics of in situ petroleum volumetrics (resources) 114
        4.3.3.2 Basics of recoverable petroleum volumetrics (reserves and resources) 115
        4.3.3.3 Subsurface uncertainty: conclusions 117

5 Aquifer analytical modeling 121

Introduction 121
5.1 Mass and energy balance 121
5.2 Van Everdingen and Hurst 122
5.3 Havlena-Odeh method 123
5.4 Cole plot for estimating aquifer contribution 124
5.5 Campbell plot for estimating aquifer contribution in an oil reservoir 126
5.6 Aquifer effectiveness in lower quality gas reservoirs 126
5.7 P/Z (Pressure/compressibility) plot 127
5.8 Solution plot for gas reservoirs 127

5.9	Carter-Tracy aquifer implementation	129
5.10	Fetkovich aquifer	130
5.11	Other aquifer models	130

6 Numerical aquifer modeling   133

Introduction   133
6.1	Required information	133
6.2	Sources of information	134
6.3	Numerical aquifer	134
6.4	Conventional gridded aquifer	136
6.5	Aquifer analytical models in reservoir simulators	136
	6.5.1 Carter-Tracy aquifer guidelines	136
	6.5.2 The Fetkovich aquifer	137
6.6	Guidelines for analytical model use	138
6.7	Using MBAL to validate an aquifer model	138
6.8	Key issues in aquifer model building	139
6.9	Estimation of aquifer–reservoir connectivity	140
6.10	Conclusions	141

7 Aquifer influx versus water injection in the Gulf of Mexico   143

Introduction   143
7.1	Aquifer description workflow	143
	7.1.1 Geophysical input to aquifer size	143
	7.1.2 Geological and petrophysical input to aquifer effectiveness	144
	7.1.3 Static uncertainty of the aquifer	146
7.2	Aquifer characterization: strength quantification	148
	7.2.1 Aquifer impact on estimated ultimate recovery (EUR)	150
	7.2.2 Dynamic uncertainty of the aquifer	152
7.3	Water injection tipping point	153
7.4	Aquifer impact on estimation of reserves and resources	154
7.5	Conclusions	154

8 Field case studies   157

Introduction   157
8.1	Field B: static modeling	157
8.2	Field B: dynamic modeling	158
8.3	Field C: static modeling	161
8.4	Field C: dynamic modeling	163

9 Applying petroleum lessons to aquifers during the energy transition 165

Introduction 165
9.1 CO_2 storage 165
    9.1.1 Geological considerations 168
        9.1.1.1 The reservoir 169
        9.1.1.2 The seal(s) 170
        9.1.1.3 Data requirements 170
        9.1.1.4 Data availability and cost 171
        9.1.1.5 The importance of economic analysis 172
        9.1.1.6 Resources available for screening analysis 172
    9.1.2 Fluid properties 172
    9.1.3 CO_2 containment modeling 177
    9.1.4 Geomechanical modeling requirements 181
    9.1.5 Geochemical modeling requirements 182
    9.1.6 Monitoring requirements 183
        9.1.6.1 A 4D example 185
    9.1.7 Sequestration capacity estimation 186
    9.1.8 The role of aquifers for CO_2 sequestration in depleted reservoirs 189
    9.1.9 Injectivity considerations 194
9.2 Natural gas storage 195
    9.2.1 Methane properties 195
    9.2.2 Gas trapping 196
    9.2.3 Best practices for natural gas storage in aquifers 197
9.3 Underground hydrogen storage (UHS) 198
9.4 Compressed air energy storage 199
9.5 Waste liquid disposal by injection into non-potable aquifers 200
9.6 Conclusions 202

Postface 205

Appendix A: Units: definitions, prefixes and conversions 207
Appendix B: Acronyms and abbreviations 209
Appendix C: Aquifer characterization and modeling checklist 213
Index 215

Preface

Subsurface professionals employed in petroleum exploration and development (E&P) spend a considerable amount of time describing and characterizing hydrocarbon reservoirs, yet the time allocated to performing the same task for aquifers is often minimal. As a result, aquifers are poorly understood and there are more uncertainties associated with aquifer modeling than in any other subject of petroleum geoscience and reservoir engineering.

Wells are seldom drilled into aquifers by design, so specific information about aquifer characteristics—porosity, permeability, thickness, net-to-gross, mechanical properties, etc.—is therefore lacking. Typical workflows infer these properties from those measured within the hydrocarbon reservoirs. Significant uncertainty is more often than not the rule when it comes to questions about the length and shape of aquifers, as well as how rock properties change with depth and distance away from the hydrocarbon reservoir.

This book aims to:

- provide a general understanding of the importance of aquifers to hydrocarbon reservoirs;
- present a consistent workflow for aquifer characterization and modeling;
- review some of the best-case studies on this topic;
- prepare guidelines for future use of aquifers for CO_2 storage.

These objectives

- are non-restrictive and allow the subsurface scientist to remain creative irrespective of disciplinary background or field of practice, whether geology or engineering;
- help define minimum deliverables and ensure a consistent approach across all fields.

The guidance provided in this book is intended to set expectations for subsurface staff on how to characterize aquifers for better prediction of aquifer performance.

This book is a critical resource for all professionals involved in the characterization and modeling of aquifers, and its insights are applicable in a variety of contexts—industry, academia, public policy, etc.

During the preparation of this book, the authors spent considerable time ensuring that all citations are properly included, and all images or tables are appropriately referenced. We would greatly appreciate readers providing us with their feedback regarding any issues related to such citations so that they can be corrected in future reprints.

Acknowledgments

We owe a debt of gratitude to our families and colleagues, whose support has made the completion of this book a reality. In particular, we would like to extend a special thanks to Dr. Mick Casey for the illustrations in Section 3.4.1 on the topic of perched water, and for his overview of anomalous water during his time working for BHP Petroleum. Chapter 6 relies in part on results obtained by Dr. Martin Cohen's aquifer modeling scoping studies. We express our admiration and gratitude for his work and generosity. Our work alongside BHP colleagues has enriched the scope and substance of our endeavor here, and many of our conversations with them have been of particular insight to the development of Chapter 7. Some of the results presented in this book, especially in Chapters 7 and 8, are based on studies carried out by a host of distinguished researchers, including Haider Rizvi, Patrick Wojciak, Jean-Sebastien Hall, Rick Gottschalk, Tony Rolland, Bellatrix Castelblanco, Mohsen Rezaveisi, Chuck Kaiser, Wing Lam, Kelly Wrobel, Tyler Foster, Jennifer Campbell, Chris Hurren, Christine Skirius and Logan Kirst, among many others. This book draws on theirs and others' knowledge, hoping to carry it forward and in turn serve as a resource for future research and developing innovations in their respective fields. We also appreciate the help we received from Silvia Clark in editing the first draft of this book.

Author biographies

Dr. M.R. Fassihi is a founder of Beyond Carbon, LLC, he is currently providing consulting services on reservoir management, project assurance and CCS with a focus on how to reduce the carbon footprint in the energy industry. His prior role was BHP Distinguished Advisor with primary responsibilities including technical assurance, competency development within the company, identification and development of emerging and white-space technology opportunities, and provision of technological advice and counsel to senior management. He was the Technical Liaison in Joint Industry Projects on CO_2 sequestration, alternative energy including geothermal, and innovative technologies for mineral extraction. Dr. Fassihi has over 40 years of experience in petroleum research, development and management of both conventional and unconventional reservoirs. His previous companies of affiliation were Arco, Amoco and BP.

Dr. Fassihi has published more than 45 peer reviewed publications and is the author of the SPE monograph *Low-Energy Processes for Unconventional Oil Recovery*. He has won numerous awards, including the SPE Student Paper Contest, PhD Division (1980), Best Paper Award at the 1988 CIM Meeting, SPE Distinguished Lecturer (2003), the international IOR Pioneer Award, Tulsa (2018) and the SPE Distinguished Member Award (2018).

J.P. Blangy is currently Practice Lead for Geophysics at BHP Petroleum, where he provides technical guidance and value insights for a broad range of subsurface matters. He was Chief Geophysicist for Hess from 2010–2016, where he was responsible for the Geophysics Department within Science & Technology as well as providing technical assurance for major projects across the enterprise. Dr. Blangy has more than 30 years of experience in the oil-and-gas industry. Prior to BHP and Hess, he spent 19 years with Amoco/BP, where he served in various management and senior technical roles across the E&P value chain, spanning from New Ventures/Access to Exploration, Appraisal, Development and Production as well as Mergers & Acquisitions. Blangy began his career with Unocal in Los Angeles (1986–1989). Blangy has authored or co-authored more than 50 papers covering a broad spectrum of topics in applied geophysics and rock mechanics. He holds a BSc. in Geophysics from the Colorado School of Mines, MSc. degrees in Exploration from Stanford University and in Petroleum Engineering from the University of Houston, as well as a Ph.D. in Rock Physics from Stanford. Dr. Blangy currently serves on the SEG's OGRC (Oil & Gas Reserves Committee) and is the recipient of several awards from the SEG.

Chapter 1

Introduction

In the petroleum literature, there are a significant number of publications that focus on detailed workflows for reservoir characterization and modeling. When it comes to characterizing aquifers however, the approaches proposed remain largely schematic due to the intent of capturing their impact through analytical or numerical modeling. The purpose of this book is to use a holistic multidisciplinary approach encompassing geology, geophysics, petrophysics and reservoir engineering for characterizing aquifers and modeling their behavior.

Aquifers are bodies of water accumulated at varying depths in underground layers of rock, sediment or soil around the world. In a petroleum-specific context, an aquifer is a zone of mobile water connected to a hydrocarbon (HC) reservoir (or reservoir intervals) at the free water level (FWL). A hydrocarbon reservoir may have multiple, possibly non-communicating aquifers, each with its own FWL. Common misconceptions concerning aquifers that we try to address in this book include:

- When it comes to aquifers, size is everything.
- Aquifers always denote increased HC reserves.
- The FWL is the same as the oil–water contact (OWC).
- Aquifer properties (salinity, porosity, permeability, etc.) are constant irrespective of depth.
- Aquifers do not contain hydrocarbons.
- Hydrostatic gradient in an aquifer is 0.433 psi/ft.
- OWC is always horizontal.
- Oil is always above the water leg.
- Permeability in an aquifer is the same as the one estimated by logs.
- Drilling an injector into an aquifer is better than drilling it into an oil leg.
- It is always preferable to use a very large grid to model an aquifer.
- You can never "outrun" an aquifer.

Aquifers play an important role in the petroleum industry because one of the key features that facilitates the maximum production and recovery of hydrocarbons is the presence and effectiveness of aquifers. On the one hand, large well-connected aquifers provide enough energy to sustain hydrocarbon production without needing to resort to improved oil recovery (IOR) operations such as waterflooding. On the other hand, reservoirs without aquifers, or with aquifers that are either small or poorly connected, have a low recovery factor, requiring additional IOR support.

Aquifer size is traditionally measured relative to the size of the hydrocarbon volume. The aquifer volume does not include any of the water in the hydrocarbon reservoir above the FWL. For example, the transition zone and interstitial water in the hydrocarbon column are not included in the aquifer volume. Hydrocarbon saturations trapped in the aquifer are not counted as part of its volume, but they can significantly influence aquifer strength. It usually takes some time for an aquifer to become active in a reservoir that has started producing from new wells. Aquifer size generally impacts reservoir pressure over the long term, as shown in Figure 1.1. Infinite aquifers maintain pressure in the reservoir indefinitely, while in reservoirs with poor or no aquifers, the pressure declines with depletion of the reservoir.

In addition to size, it is important to consider aquifer effectiveness and location relative to that of the reservoir. High connectivity or permeability between a well producing hydrocarbons and an aquifer enhances the aquifer's energy impact. Different levels of connectivity provide different levels of pressure support to the reservoir's energy system. Because of their proximity to the reservoir, bottom water drive aquifers have a much higher immediate impact on recovery than do edge water aquifers. The location and way in which aquifers interact with the hydrocarbon reservoir play a significant role during early production, as shown in Figure 1.1. A large or infinite-acting aquifer is likely to take a long time to make its influence felt. When poor aquifer performance is suspected or detected, it is preferably managed by using water injection wells that provide energy support to the reservoir system.

An aquifer's impact is also dependent upon the fluids present within the reservoir. For instance, if the hydrocarbon is in a liquid state instead of gaseous, the compressibility of the system is low, meaning that pressure support is transmitted to the reservoir relatively quickly. As a result, even relatively small aquifers on the order of two-to-one or three-to-one (compared to the hydrocarbon volume) can provide significant energy in the short term. Water injection may not be required immediately or may be delayed for medium-size aquifers (which are

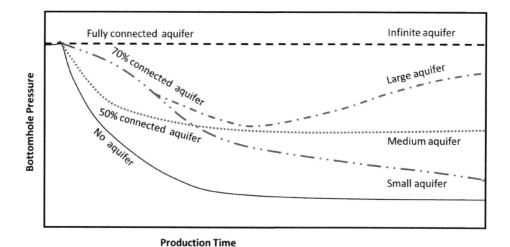

Figure 1.1 Impact of an aquifer on a hydrocarbon producer.

typically on the order of five-to-one to ten-to-one in size) that act relatively quickly. The presence of aquifers is beneficial in these instances, because the reservoir's energy can be sustained for a longer period than if there were no aquifers. Conversely, for high compressibility reservoirs such as those containing dry gas, the impact of a large aquifer is usually detrimental because of gas-trapping.

There are more uncertainties associated with aquifer modeling than in any other area of reservoir engineering. It is uncommon for wells to be purposefully drilled into aquifers, and therefore information is often lacking regarding aquifer-specific properties such as porosity, permeability, thickness, net-to-gross and the mechanical properties of the rock. These properties are usually inferred from measurements taken within the hydrocarbon reservoir. Relying on these local measurements and on a poor sampling of the reservoir system generates uncertainties related to the aquifer's length and shape, as well as variation of rock properties with depth and distance within the aquifer. A "one size fits all" approach to aquifer characterization has limitations which we cover in the various chapters of this book.

Unlike other, geoscience-based publications on aquifer characterization that focus on aspects such as hydrology, this book approaches aquifer characterization from the perspective of multidisciplinary lessons learned from the petroleum industry. The methodologies discussed in this book and illustrated through a few case studies are intended to show how to derive additional information on aquifers and thus reduce the uncertainty surrounding the parameters needed for properly modeling hydrocarbon production and estimating recovery. In order to help those who embark on characterizing and modeling aquifers, we provide a modeling checklist and a series of guidelines that will assist them in their studies.

This book is organized into nine chapters. Chapters 2 through 4 of this book discuss the types of data and methods used for aquifer detection, identification and characterization. Aquifer static modeling techniques such as basin modeling, touchstone modeling and petrophysical modeling are discussed in Chapter 5. Our subsequent discussion of dynamic aquifer modeling in Chapter 6 takes both analytical and numerical models into consideration. When developing a new field, it is important to determine whether, and/or when, water injection might be needed. This aspect is discussed in Chapter 7. In Chapter 8, we turn to three in-depth field case studies that illustrate some of the modeling techniques covered in earlier chapters. In light of rising interest in carbon capture and sequestration (CCS) technologies, saline aquifers deserve increased attention as potential sites for future CO_2 storage. In Chapter 9, we show that the workflow presented in Chapters 2 through 6 is also applicable to the characterization of saline aquifers and to their evaluation as suitable CO_2 storage sites. Our suggested checklist for aquifer modeling, a glossary and unit conversions are provided in the appendices.

Chapter 2

Aquifer presence and identification

INTRODUCTION

Geoscientists typically think of reservoir systems in terms of depositional environments within which depositional processes occur through geological time at regional scales, long before any hydrocarbons (HC) are generated. Reservoir systems are therefore expected to extend beyond HC contacts, from structurally elevated locations, which may contain oil or gas, into deeper synclinal locations constituting aquifers. In a working petroleum system, hydrocarbons are expelled from the source rock, and partially displace water within a geological trap or container. Gravity segregation and fluid equilibration processes generally result in the presence of HC up-dip and water down-dip within such traps. As a result, all hydrocarbon reservoirs are expected to have some form of associated aquifer. What remains to be determined is the extent, quality and effectiveness of these aquifers.

Typical workflows infer the presence of an aquifer from seismic interpretation and mapping. However, geophysicists and geologists do not always have the right data to map the extent of those aquifers. In the absence of concerted efforts in both data gathering and interpretation dedicated to the assessment of aquifers, one is forced to contend with great uncertainty as to the volume of the aquifers and their shape or distribution.

This chapter discusses key aspects of aquifer description and characterization derived from the interpretation of data from the various scales of investigation of three fundamental subsurface disciplines—geology, geophysics and petrophysics. This chapter enables readers to:

- Understand the kinds of data that can be used to characterize aquifers;
- gain insight into why aquifers are usually more difficult to detect than HC reservoirs.
- Become familiar with an example of seismic workflow that is applicable to soft (or low acoustic impedance) reservoirs and that can be used to detect and map aquifers with relative precision and accuracy.

The chapter is not meant to be an exhaustive review of the workflows employed by all three disciplines. Instead, it presents the essential concepts that form the technical basis for the construction of realistic earth models, a topic that will be addressed in Chapters 4 and 5.

2.1 DEFINITIONS

Aquifers are underground bodies of water-bearing, permeable rock that can be found in a variety of geological environments, such as clastic sediments of various grain sizes (sands, silts, gravel), carbonates, or fractured host rock (crystalline or igneous).

The discipline of hydrogeology focuses on the characterization of shallow aquifers, as well as the understanding of the flow of water in and out of those aquifers, as part of a system of groundwater charge/discharge within an overall hydrologic cycle. Figure 2.1 illustrates some of the key concepts of water discharge and recharge into various types of aquifers. Under natural equilibrium, the amount of recharge to the water table equals the amount of discharge from streams and/or runoff. Pumping disrupts the equilibrium, thus removing water from storage. Recharge may take days to years or may not happen at all, depending on the water budget.

There are multiple subdisciplines within the field, such as hydrogeophysics, which interfaces with hydrogeology, geochemistry, soil mechanics and civil engineering.

To note, aquifers do not necessarily contain fresh water (more on this in Section 2.3.2.3). Because they often originate from the interaction of meteoric water recharge and stream water discharge (Figure 2.1), aquifers can be found in many places. Meteoric water derives from precipitation (rain and snow) and includes rivers, lakes and ice/glacier meltwater. A critical concept for understanding aquifers is the ***water table***, defined as an underground potentiometric surface in an unconfined system where the pressure head of water equals atmospheric pressure, resulting in a state of equilibrium for flow. Hydrologists define surfaces of constant head, useful for visualizing where the water is flowing. The pressure head (or static head) is the height of a water column, measured in units of distance from the water table. The depth of the water table is controlled locally by capillary forces in the soil.

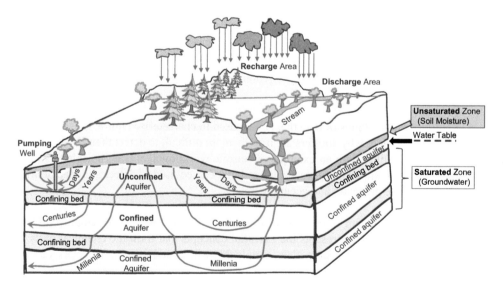

Figure 2.1 Schematic principles of the hydrologic cycle. (Adapted from Freeze and Cherry 1979 and from Figure 3 in Circular 1139 of the United States Geological Survey.)

Above the water table, the pressure head is negative, so water is under suction toward the surface. Within this shallow zone, water fills the soil's pore space in a discontinuous, often patchy fashion. This is a region of infiltration from the surface and the water is kept in situ by adhesive forces occurring at the scale of grain contacts. In other words, the depth to the water table is controlled locally by the capillary forces within the soil. This area between the earth's surface and the water table is known as the *vadose zone*, and it contains **aquifers in an *unsaturated state.*** On the other hand, below the water table, the pressure head is positive, so water flows down and away from the vadose zone, striving for pressure equilibrium. In this area, known as the *phreatic zone*, all pore space is fully saturated with water. This book concerns itself mostly with **saturated aquifers**.

Another important distinction is that between ***confined*** and ***unconfined aquifers***. Unconfined aquifers are in communication with the Earth's surface, and therefore they are geological bodies where groundwater can be recharged directly. In contrast, confined aquifers are overlain by a sealing surface that prevents flow across it (Figure 2.1). Confined aquifers generally require a minimum depth of burial for the seal above the aquifer to be effective and to prevent flow across it toward the Earth's surface.

Aquifers can occur in all types of rocks—sedimentary, igneous, metamorphic—though the last two require a significant amount of fracturing to be effective. The best aquifers are usually encountered in sedimentary rocks, with both clastics and carbonates able to form excellent aquifers. Some of the most striking examples of aquifers are contained in karsted carbonates, which can accommodate large-scale underground rivers whose extent is rarely visible to explorers.

An important area of study in hydrogeology concerns the **dynamic equilibrium** of aquifers (Konikow and Bredehoeft 2020). First discussed by Theis (1940), the principle of dynamic equilibrium states that all water discharged by wells must be balanced by water loss elsewhere in the system. Under natural conditions, recharge at the water table is equal to discharge at the stream. If water wells are introduced into the system, the water lost to those producing wells must come from either storage in the aquifer or from capture of additional groundwater. The capture of groundwater requires an increase in recharge to avoid any depletion of the aquifer. Because groundwater capture is highly variable both in time and in space, any constraint causing insufficient recharge of the aquifer will disrupt its equilibrium and lead to its eventual depletion, which is an unsustainable outcome.

Understanding the extent of underground aquifers, their potential volumes and state of equilibrium requires a thorough exercise in mapping using observations obtained from the surface and the subsurface.

2.2 DATA AVAILABILITY, COST AND SCHEDULE

A significant amount of subsurface data can be obtained from government agencies such as USGS (United States Geological Survey), from scientific organizations like the International Ocean Discovery Program (IODP), or from the oil and gas industry (O&G). Overall, O&G has been the leading source of funding for exploration drilling and therefore has accrued the largest store of offshore data available. Consequently, the insights in this book are derived in large part from industry-sourced offshore data.

In this context, the two main industry-specific types of data relevant to locating and understanding aquifer systems in the subsurface are ***direct measurement*** through wells and *remote sensing* through various physical properties measured from a distance. Sections 3 and 4 of this chapter expand upon the types of well data and the types of remote sensing data that can be used to assess reservoir rocks and their aquifers.

When considering offshore O&G, it is helpful to group fields into two broad categories that reflect their maturity, as well as the type and amount of data available from those fields: fields that are in pre-production, and fields that are under development and production.

2.2.1 Fields in the late exploration or appraisal phase

Reservoir rock and fluid properties retain considerable uncertainty for fields in the late exploration or appraisal phase. At this stage, a field may be covered by 3D seismic and may contain a few well penetrations. To infer the presence, extent and type of the aquifer from data available at this stage, the properties being tracked are typically the following:

- rock type and quality
- reservoir properties: porosity, permeability, HC saturation, compressibility
- fluid properties from early pressure-volume-temperature (PVT) analyses, including API gravity, gas-to-oil ratio (GOR), oil formation volume factor (B_o), gas formation volume factor (B_g), water saturation (S_w) under initial conditions, as well as effective pressure under initial conditions
- seismic elastic properties: Acoustic Impedance (AI), Shear Impedance (SI), Gradient Impedance (GI), or Lithology Impedance (LI) and Fluid Impedance (FI), where geophysicists use the Fluid Application of Geophysics (FLAG) correlations (Han and Batzle 2014) to assess the impact of fluid properties on the seismic response

The use of pre-stack seismic amplitudes, once calibrated to wells, has proven particularly useful in assessing the types of fluids within certain reservoirs via remote sensing.

2.2.2 Fields under development and/or production

For fields under development and/or production, the properties of the reservoir rocks are, in general, fairly well understood and are quantified at various locations using data from producer and/or injector wells. At this stage, the field may be covered by 4D seismic and may also have a thoroughly documented production history—as is increasingly common given oil field digitization. If the latter is indeed the case, an integrated interpretation of the joint surveillance data (from the wells and the 4D) can be used to assess the movement of fluids during the recovery process and to gauge the uncertainty around remaining recoverable volumes. In addition to the properties listed in Section 2.2.1, the data typically available for fields in the development and/or production phase can be used to determine the following reservoir properties:

- compartmentalization, or structural complexity due to faulting and/or stratigraphic variation
- connectivity (as seen through the compartmentalization of the reservoir)
- advanced rock properties: compaction and diagenesis (cementation, clay diagenesis, etc.)

- properties of the fluid compartments, including pressure, saturation, water salinity (if water has been sampled)
- recovery mechanisms [i.e. primary depletion with or without aquifer support, gas cap expansion, water flooding, field under enhanced oil recovery (EOR) method]
- remaining fluid distribution as seen through S_w, oil saturation (S_o), gas saturation (S_g)
- effective pressure of the wetting phase
- overburden properties (e.g. subsidence, stress arching)

Generally speaking, more sophisticated data (e.g. detailed fluid sampling and/or seismic-based reservoir characterization) require more time (from one to several years) and expense than reconnaissance data. Shallow and onshore data are faster and/or cheaper to acquire than offshore and deep data. Derived data such as core data and laboratory analyses require wells to be drilled, and drilling is usually the major source of cost. A schematic summary of typical cost-schedule relationships is shown below in Figure 2.2.

The following sections contain further discussion of the data types acquired through wells and via remote sensing.

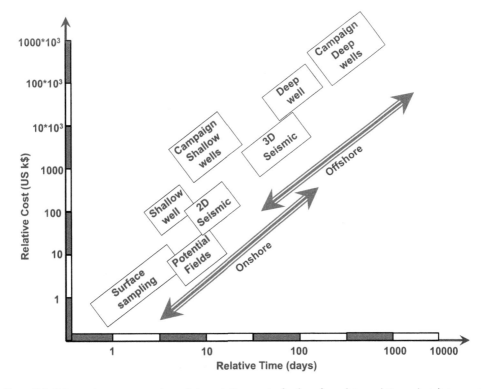

Figure 2.2 Schematic representation of the relative cost of subsurface data and its analysis/interpretation. Indicative costs and timelines are shown for reference.

2.3 DETECTION METHODS: DIRECT DETECTION

2.3.1 Surface sampling and mapping

Aquifers sometimes outcrop, in which case their extent can be effectively inferred through geological mapping and direct water sampling from creeks, springs, etc. Surface sampling is feasible only where geological formations breach the Earth's surface (i.e. mostly onshore). An alternative to surface sampling is to drill shallow water wells in the very-near surface (down to depths of several tens or even hundreds of meters) and that is a time-tested, reliable technique for accessing aquifers. However, owing to the fact that the majority of the planet lies either under bodies of water or below ground surface cover, and due to the costs involved in drilling, indirect (i.e. remote sensing) methods are required to systematically detect and map potential aquifers everywhere across the Earth's near surface.

2.3.2 In situ sampling or well-based methods

Wells are the main source of accurate information about underground aquifers, because they enable the direct physical sampling of fluids and rocks in the subsurface. Drilling for water goes back to at least 8400 BC, as documented by records of early Neolithic wells dug into limestone on the island of Cyprus in the East Mediterranean (Peltenburg 2012). The technology for constructing wells (i.e. drilling and completions) has made significant advances, and wells will continue to be the main source for characterizing the quality both of the water and of the formation housing it in the subsurface.

Once a well is drilled, there are a number of ways to analyze the in situ properties of the aquifer. The key methods rely on well-logs, pressure, temperature measurements and direct rock and fluid sampling. Well measurements are typically impacted by "damage" from the drilling process. Nevertheless, subsurface samples of rock and fluids are often "preserved" as much as possible as they are retrieved to the surface for analysis and for the calibration of well logs.

2.3.2.1 Well logs

Well logs are measurements made with a sonde that is lowered inside a wellbore. Different types of well logs are available, depending on the type of property being tracked. Well logs typically measure a property within a distance ranging from several centimeters (cm) up to 10 meters (m) away from the well. The main logs recorded in the industry are gamma ray, resistivity, sonic, density and nuclear magnetic resonance (NMR). Other spectral logs that help identify mineralogic composition from base elements are sometimes recorded as well. The application and interpretation of some of these logs is described in more detail in Chapter 3, in the section on petrophysics.

2.3.2.2 Pressure and temperature

Two important in situ properties that are difficult to predict with accuracy from the surface are pressure and temperature, so it is best to measure them directly whenever possible. Several methods are commercially available to determine downhole pressure and temperature.

Temperature logs are obtained from values equilibrated between formation temperature and drilling fluid temperature. Temperature probes are lowered in the borehole and typically record a maximum temperature value. They are considered to have an accuracy of at best 0.5 °C (Prensky 1992).

Pressure measurements are obtained using various logging tools such as repeat formation testers (RFT), modular dynamic testers (MDT) and drill stem testers (DST). Their readings provide total formation pressure measured from sea level, and their accuracy is generally around 0.5–1 psi (35–70 millibars). Key attributes of each of the three tools include:

- DST readings, which provide the most reliable source of downhole pressure information, typically require packers placed above and below the formation of interest so as to isolate the zone. DSTs can be obtained both in flowing and non-flowing (or shut-in) mode. DST tools tend to sample a relatively large formation interval, ranging from several feet to tens of feet, and allow for the testing of multiple flow units.
- The RFT comes in contact with the formation, taking fluid samples from different depths at small intervals. Proper RFT functioning requires a good seal between tool and formation.
- The MDT is similar to a mini-DST except that it is run on wireline, where the information is stored, so it does not require a drill stem. However, the MDT only samples a small volume located around the tool's pressure probe.

For a more thorough discussion on geopressure measurements and their interpretation, please see Dutta et al. (2021) and Flemings (2021).

2.3.2.3 Rock and fluid samples

Rock samples

Listed in order of acquisition time and cost, rock samples come in the form of drilling cuttings, percussion sidewall cores, rotary sidewall cores and whole cores.

- Cuttings are typically broken or pulverized, and therefore not suitable for detailed laboratory analyses, except if occurring in larger fragments due to unstable wellbore conditions.
- Percussion sidewall cores are recovered by impact from a hollow projectile sent into the formation and are typically only used for mineral composition analysis.
- Rotary sidewall cores attempt to preserve the fabric of the rock as they are drilled though the borehole wall so that they can be used for routine core analyses (RCAL), including porosity and permeability measurements.
- Whole cores are recovered as large sections of the formation measuring up to 30 m (90 ft) each. The cores are handled carefully and are suitable for special core analyses (SCAL), including wettability and relative permeability.

Fluid samples

Aquifers display a wide range of water quality, a critical factor that determines their use by humans. Aquifers can be a prized source of fresh water when they contain low concentrations

12 Integrated Aquifer Characterization and Modeling

of dissolved salts and total dissolved solids (TDS). However, most water encountered in a natural state requires some form of treatment prior to human consumption. The best way to assess the quality of water in an aquifer is to sample and analyze it in the laboratory.

Salinity, typically measured in parts per thousand (ppt) and sometimes in thousands of parts per million (k ppm), is a key factor in assessing water quality, as summarized in Figure 2.3. The Figure illustrates the most common categories used to classify water quality: fresh water

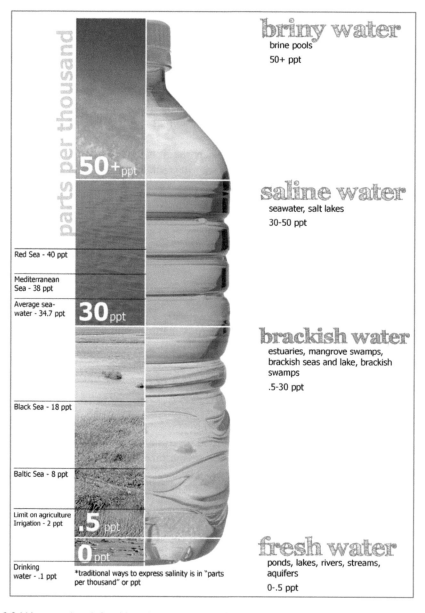

Figure 2.3 Water quality defined by salinity and uses. (Summerlin 2011.)

Aquifer presence and identification   13

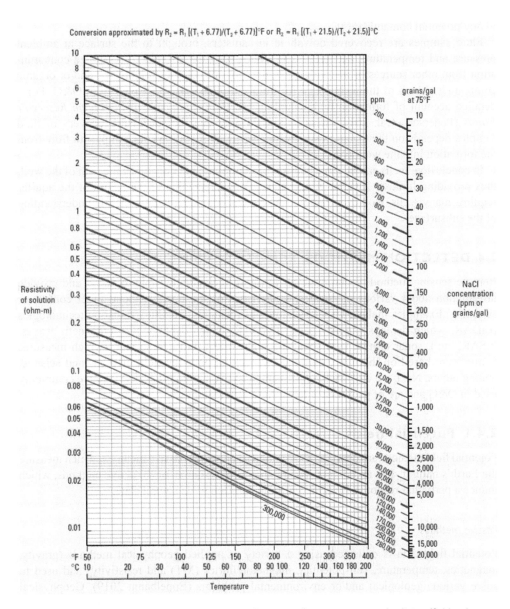

Figure 2.4 Resistivity of a saturated solution as a function of temperature and salinity. (Schlumberger 2013.)

(<0.5 k ppm), brackish water (0.5 k – 30 k ppm), saline water (30 k – 50 k ppm) and brine (>50 k ppm).

Approximate subsurface water salinity can be estimated from measurements of formation water resistivity recorded in well logs (see Figure 2.4). However, in order to assess the quality of an aquifer, it is always best to sample the water directly so as to obtain accurate measurements not only of salinity, but of the overall chemical composition and, in particular, of any potential contaminants.

Fluid samples are recovered downhole in canisters, brought to the surface at ambient pressure and temperature, and processed with great care in order to minimize contamination from other sources or intervals, including possible dissolved inorganic and/or organic compounds. Much of the technology for fluid sampling can be adapted from O&G. For a detailed account of formation fluid sampling see *Phase Behavior of Petroleum Reservoir Fluids* (Pederson et al. 2014). Readers should note that the successful recovery of fluid samples depends on the extent to which rock permeability allows for samples to flow from the formation into the wellbore.

In conclusion, wells enable direct subsurface sampling, but only along the path of the well, thus providing limited, 1D data. A regional, 3D understanding of the nature of the aquifer requires many wells. However, remote sensing can provide a large-scale 3D understanding of the subsurface.

2.4 DETECTION METHODS: REMOTE SENSING

Remote sensing methods incorporate all forms of technologies able to detect and provide information about various physical characteristics of the Earth without direct contact or sampling. Examples include visual technologies (cameras, satellite), radiation technologies (LiDAR, which deploys a pulsed laser, RADAR, which deploys electromagnetic waves, InSAR, etc.), sonar (which uses sound through water), infrared sensing (which measures temperature based on thermal emissions), as well as potential field methods and seismic, among others. Next, we will briefly discuss the last two methods because they are commonly used in O&G and are of particular interest for aquifer studies.

2.4.1 Potential fields

Potential field methods are commonly classified as either (a) passive methods, which measure the Earth's natural state and do not require an emitting source, or (b) active methods, which induce a particular response from the Earth, in addition to its natural state.

Passive methods

Potential field data is collected using a variety of applied geophysical methods (gravity, magnetics, temperature, self-potential, magnetotellurics (MT) and resistivity) and used to solve various geological and/or environmental problems (Eppelbaum 2019). Geophysical potential methods (excluding resistivity) record the Earth's natural fields and are therefore referred to as passive methods. For practical purposes, potential fields are assumed to be time independent, at least during the acquisition time. Examples of potential methods include gravimetry, magnetometry, as well as some electric resistivity methods and MT.

Because they are relatively inexpensive, gravity and magnetic surveys are often conducted regionally over vast areas of land and sea. MT methods record the horizontal component of the Earth's naturally occurring electro-magnetic field at the seafloor. They enable the measurement of electrical resistivity from the seafloor down to various subsurface target depths, which are adjusted through the frequency content of the signal.

Active methods

Potential field data is sometimes augmented by using active sources to induce a response that is measured using similar approaches to passive field data.

For example, controlled source electromagnetic data (CSEM) uses an electrical dipole as an emitting source in order to induce an electromagnetic field. Like MT, it allows for the measurement of the resistivity of the Earth at depths varying from several meters to a few kilometers, depending on the frequency band and on the source-receiver offset.

2.4.2 Seismic

Seismic interpretation is a central function of the geoscientist. During seismic interpretation, much attention is focused on mapping and understanding the reservoir, its depositional environment and its structural history. Many of the types of reservoirs that are mapped have a lower acoustic impedance (AI) than the background rock (i.e. they are softer than their surroundings), due either to present/preserved porosity, HC, or overpressure. As all three factors are known to lower AI, the reservoir section often manifests as a "sweet spot" region of lower AI, which can be identified as a seismic amplitude anomaly on stacked seismic sections. The enhancement of the hydrocarbon-filled sections for these soft reservoirs usually requires important, detailed seismic processing steps whose description is beyond the scope of this book.

Unfortunately, the common O&G practice of interpreting "anomalous" seismic amplitudes to identify the limits of HC reservoirs often breaks down within the aquifer section of the same reservoirs, because aquifers typically have AIs that are similar to their surroundings. In the case of reservoirs and/or aquifers with low AI contrast, a rock-physics-based approach using pre-stack characteristics can sometimes be adopted to characterize their properties, as discussed in Section 2.4.2.2.

The advantage of seismic methods stems from their relatively high vertical and lateral resolution, which is typically on the order of 10–50 m. The main disadvantage of seismic methods is their high cost and relatively long cycle-time from acquisition to interpretation. A typical seismic dataset might require 2–6 months to design and permit, another 2–4 months to acquire and 6–12 months to process, migrate and interpret.

2.4.2.1 Seismic data characteristics

As a seismic compressional or P-wave propagates through the Earth, it is transmitted through layers of rock and encounters interfaces, where some of its energy is reflected back, some continues to be transmitted downward as a P-wave, and some is converted into other waves (i.e. shear or S-waves). This wave portioning phenomenon (shown schematically in Figure 2.5) has been described systematically, and its physics is approximated by a series of equations called the Zoeppritz equations (Zoeppritz 1919).

Seismic is recorded in time as a waveform that includes signal from P-waves (incorporating a direct wave, reflected waves and refracted waves), S-waves, arrivals from other waves, and noise. A seismic trace is a record of ground motion as a function of time. Three-component geophones record ground oscillations in X, Y and Z, while hydrophones record a pressure pulse (in water), as a function of time recorded from the surface. At any recorded time t, the key characteristics of a seismic trace are ***velocity***, ***amplitude***, ***frequency*** and ***phase***.

- **Seismic velocity** is a transmitted property measured as a seismic wavefront that propagates through the Earth. It is the rate at which a seismic wave travels through a medium, expressed by the ratio of distance and travel time. Velocity can be measured as an instantaneous quantity at a certain point in time, or as an average for an interval. Because velocity is transmitted, it exhibits a relatively low resolution. Recent advances in seismic acquisition and processing leveraging newly developed full waveform inversion (FWI) algorithms are substantially improving the accuracy and reliability of our subsurface velocity models.
- **Seismic amplitude** is the deviation of a passing wave from zero crossing measured at a given point in time. It can be measured instantaneously or as an average over a certain interval.
- **Seismic frequency** is the number of cycles per second that a particular seismic wave exhibits. This quantity can also be measured both instantaneously and as an average over a certain interval. Typical seismic waves contain a range of frequencies, which is assessed by their frequency spectrum. It has been shown that instantaneous frequency relates to the centroid of the frequency spectrum of the seismic wavelet. The dominant frequency of a seismic wave, f (in 1/s) is related to its velocity (V in m/s) and wavelength ($\lambda\, in\, m$):

$$f\left(\frac{1}{s}\right) = \frac{V\left(\frac{m}{s}\right)}{\lambda(m)} \quad (2.1)$$

- **Seismic phase** is a measure of a wavelet's true phase. This quantity can be measured instantaneously or as an average over a certain interval. The rate of change of the instantaneous phase is equal to the instantaneous frequency.

A large number of additional seismic attributes can be derived from the four characteristics described above. In fact, complex seismic attributes have been the subject of considerable research. For a good classification of seismic attributes see Taner et al. (1979), and for a good overview as well as a historical perspective see Chopra and Marfurt (2005).

There are other types of seismic attributes, such as intrinsic attenuation—caused by energy dissipation as seismic waves propagate through the Earth. Geophysicists have historically limited their analysis of seismic data to seismic reflections (P and S waves), refractions and noise, but the practice is currently evolving with the recent development of FWI technologies. So far, and due to computational requirements, we can state that O&G has not fully leveraged a complete understanding of seismic propagation because it has limited analysis to reflected waves.

2.4.2.2 Seismic data type, availability and quality considerations

There are two main categories of seismic waves, i.e. body waves and surface waves. Body waves travel through the Earth's inner layers and consist of P-waves and S-waves. Surface waves are "trapped" and travel along the Earth's surface; they consist of Rayleigh and Love waves.

O&G has relied upon, and is further developing, the use of seismic reflections. Reflections occur at the site of contrasts in impedance (the product of velocity times density) within the subsurface and tend to display an acceptable resolution. Contrasts in impedance can be due to stratigraphic reasons (changes in facies, unconformities, etc.) or to structural reasons (e.g. faults, areas of tectonic deformation, uplift causing unconformities). As mentioned earlier, recent advances in our burgeoning understanding of the full seismic waveform and its propagation from the surface downwards are expected to further improve the resolution of seismic data and our ability to decipher the seismic signal.

When considering seismic reflections, we distinguish between stacked data versus pre-stack data. The latter requires longer preparation/processing times and greater attention to detail on account of its lower signal-to-noise ratio (S/N). However, the benefits of pre-stack data include the potential recovery of both P-waves and S-waves.

Stacked data. Stacking is the process by which data is summed up from a common reflection point. Because reservoir signal is often consistent trace-to-trace and seismic noise is often random (except for some types of coherent noise, e.g. seismic multiples), stacking greatly enhances the quality or S/N of the seismic.

The stacked section is a proxy for an AI section, and can often be thought of as a seismic section responding to the relative "hardness" of the rock.

Pre-stack data. Pre-stack data leverages the fact that waves convert at rock interfaces. For example, as shown in Figure 2.5, when a down-going P-wave intersects an interface with a given impedance contrast at an angle θ_1, it is both transmitted downwards as a P-wave (at an angle θ_2) and reflected upwards (at an angle θ'_1), as well as partially converted into a P-S wave that is simultaneously reflected upwards (at an angle δ_1) and transmitted downwards (at an angle δ_2). The geometric relationships for the P- and S-wave pathways follow Snell's law. The physics of elastic energy conversion at the lithology interface described by the Zoeppritz equations (1919) accounts for the conversion of amplitudes from one elastic wave mode to another.

Seismic processing workflows used in pre-stack data analysis aim to recover amplitude and angle relationships from the data through amplitude-versus-angle (AVA) analyses. The main benefit afforded by pre-stack data consists in the quantitative interpretation (QI) of this conversion derived from pre-stack seismic.

Pre-stack data interpretation is often worth the extra time, money and effort because it enables the breakdown of the seismic signal into components related to the rock itself and components related to the fluids contained within the rock. Geophysicists working with pre-stack data often create a *seismic lithology* volume and a *seismic fluid* volume. Four main types of AVA (Amplitude versus Angle) classes (I through IV) have been recognized (Rutherford and Williams 1989; Castagna et al. 1998), and elegant formulations have been devised to derive rock and fluid property estimates by introducing elastic impedance (EI) concepts (Connolly 1999) as well as extended elastic impedance (EEI) formulations (Whitcombe et al. 2000).

Pre-stack data analysis unlocks a more systematic and in-depth use of raw (i.e. "as recorded") seismic data and its amplitudes. However, pre-stack data suffers from a much lower S/N, so the interpreter needs to understand both the signal and the noise that arise from seismic propagation in the subsurface, as well as the changes that occur in the seismic wavelet through time. In other words, pre-stack data requires signal processing workflows that "preserve" amplitudes. Some of these workflows include data conditioning steps that enhance the quality of the pre-stack gathers. Because of their importance, such steps should involve specialist input from a processing geophysicist and/or a quantitative seismic interpreter.

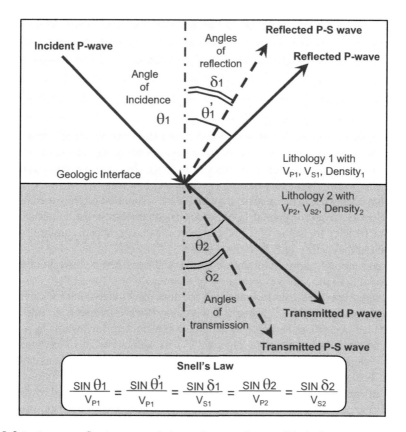

Figure 2.5 Seismic wave reflection, transmission and conversion at a lithologic

Most pre-stack data conditioning workflows aim to honor relative amplitude-preserving steps such as:

- scaling, which can include Q compensation for phase and/or amplitude
- alignment: de-wobble via higher-order normal moveout correction (NMO) or Dave Hale algorithm
- noise attenuation
- frequency balancing, which can include wavelet normalization

The *elastic impedance* method (Connolly 1999) derives the impedance equivalent to an angle-dependent reflectivity R (θ), where θ is the angle of incidence, through a mathematical transformation involving parameter χ. Quantitative geophysicists have found that for most turbidite reservoirs that have not yet undergone significant diagenesis, χ takes on values of approximately 0, 20 and 90 for the seismic stack, seismic fluid and seismic lithology volumes, respectively. EEI, a generalization of Connolly's method, was shown by Whitcombe et al. (2000) as capable of enabling the estimation of several elastic parameters, among which bulk and shear moduli, which are sensitive to pore fluid type and to lithology. See Section 2.5.2.2

for an example of seismic inversion using EEI. EEI workflows can be applied in conjunction with well data for calibration to "absolute" rock properties (preferred option) or as self-calibrating processes for estimating "relative" rock properties.

2.4.2.3 Seismic resolution and seismic detection from reflections

Both seismic detectability and seismic resolution can be assessed once the amplitude and phase spectra of the seismic wavelet have been estimated.

Seismic resolution

Vertical resolution is the ability to differentiate the top from the bottom of a layer. It can be shown to be exactly ¼ of the wavelength of a perfectly symmetric Ricker wavelet, and around ¼ of the wavelength of the peak frequency contained in the seismic. The wavelength, λ (in ft), is the average P-wave velocity (in ft/s) divided by the peak frequency, f (in 1/s):

$$\lambda = \frac{V}{f} \tag{2.2}$$

Below seismic tuning thickness, which is approximately ¼ of a wavelength, seismic beds are considered "thin beds," where top and bottom cannot be differentiated (Widess 1973), so seismic interpreters resort to approximating bed thickness using a seismic amplitude approach.

Since resolution is a function of wavelength, shallow seismic displays higher resolution than deeper seismic. Considering a shallow horizon with P-velocity of 2,000 m/s and a peak frequency of 50 Hertz (Hz), the wavelength is 40 m and the limit of resolution is ~10 m. A deep horizon with a P-wave velocity of 3,500 m/s and a peak frequency of 20 Hz has a resolution limit of ~45 m (43.75 m).

Lateral resolution is the minimum horizontal width of a feature that can be imaged. Seismic reflected waves interfere where their travel paths are less than half a wavelength apart, and the portion of the reflecting surface involved in the interference is called the First Fresnel zone. As with vertical resolution, researchers have shown that the limit of horizontal resolution is defined as approximately ¼ of the wavelength from a migrated seismic section. Seismic migration not only allows for the proper positioning of seismic events in 3D, but it also improves lateral resolution.

As in the discussion on vertical resolution, when considering a shallow horizon with P-velocity of 2000 m/s and a peak frequency of 50 Hz, the lateral resolution limit is ~10 m. A deep horizon with a P-wave velocity of 4000 m/s and a peak frequency of 20 Hz has a resolution limit of ~50 m.

Seismic detectability

Seismic detectability is the ability to identify the existence of a bed without necessarily being able to decipher its geometry. Detectability is a function of the frequency bandwidth and is typically on the order of one tenth of a wavelength for a perfectly symmetric Ricker wavelet. Understandably, the detection limit is a function of S/N.

2.4.2.4 Seismic observations

Seismic observations are impacted both by the overall *quality/fitness of the data* (a characteristic of the data itself) and by the ensuing *seismic expression* (a characteristic of the imaged object). Our discussion of resolution and detectability shows that objects with small dimensions (either vertically—e.g. very thin beds, or horizontally—e.g., narrow features) cannot be imaged unless they offset bigger features. For instance, faults with significant displacements are routinely imaged even though the width of a fault zone can be quite narrow.

If one wishes to interpret seismic "quantitatively"—i.e. beyond geometric observations—for characterizing the rock and fluid in the subsurface, two steps must be considered:

(1) *Data quality assessment*: Seismic data processing history and aptness/fitness for QI

It is important to establish a control on seismic data quality in order to evaluate legacy seismic processing workflows that were used as part of the history of the data (a look-back approach), as well as to form an opinion on the adequacy of the data for future processing projects (a look-forward approach). This is particularly important in the case of pre-stack data, because seismic QI is an integrated and iterative process that is highly data-dependent as well as sensitive to signal processing. Scanning through the seismic data in various inline, crossline and time/depth-slice directions enables both the evaluation of seismic lateral consistency and the detection of any organized noise, often best identified on seismic pre-stack gathers (e.g. multiples, PS-conversions, etc.). As such, all quantitative seismic observations should start with a review of:

- The variability in the stacked seismic data at the project scale, e.g., by using maps of derived-seismic attributes over sizable geologic intervals. These maps can be based on seismic amplitudes, coherency, frequency, etc.
- More detailed extractions of various seismic attributes at the target/reservoir level, e.g. amplitude/background ratios (A/B), root-mean-square (RMS) amplitude, etc., so as to (a) allow for a timely identification of overburden or surface imprints, and (b) reveal artifacts inherited from seismic acquisition, illumination or merges in various surveys. If a significant amount of lateral variability is detected upon review, and if it is independent of potential artifacts due to (a) or (b), check that the observations of lateral changes in the data are consistent with what might be expected from a geological context for the area.
- pre-stack data quality (through inspection of seismic gathers) across the seismic survey.

(2) *Seismic expressions*: Key seismic observations at the target/segment level

During exploration and appraisal, the calculation of in-place fluid volumes is typically identified and aggregated based on the concept of geological segments. A *segment* is a connected volume of fluid in the subsurface, either proposed (if undrilled) or known (if drilled), that can be characterized by a single volumetric distribution function and a single geological risk. Risk is set to 1 if the segment is drilled and forms a known accumulation. Because the segment forms the basis for calculating fluid volumes, detailed discussions of size and probability should be held at the individual segment level. This applies to both HC and aquifer segments.

In order to define an interpretable body or segment, it is important to recognize that there are numerous reasons for seismic attribute observations to deviate from the background. It is convenient to group these factors into variations in:

- rock properties (lithology, porosity, cementation, diagenesis, etc.)
- fluid properties (gas, oil, water, saturation)
- in situ conditions (e.g. temperature, pressure)

When identifying and tracking a particular segment while interpreting seismic, it is important to decipher the signal contained within the seismic (amplitude, phase, frequency and velocity as discussed above) in order to best understand the subsurface distribution of fluids, lithology, reservoir thickness, clay content, porosity, net-to-gross (NTG), effective stress and pore pressure (see Chapter 3 for examples). It is a good idea for the overall seismic interpretation of a segment in terms of size, shape and internal character to be consistent with geological expectations and/or known analogs.

Earlier in this chapter, we emphasized that well-based mapping methods are more reliable and accurate than remote sensing methods, because wells sample the subsurface directly at defined locations and thus enable a direct measurement of the rock and fluids at subsurface pressure and temperature conditions. On the other hand, remote sensing methods allow for a complete yet indirect 3D sampling of the subsurface, and usually unlock an understanding of possible distributions of rocks and fluids below the ground. The maximum amount of reliable information about the subsurface is derived by combining and integrating data from both wells and remote sensing.

2.5 EXAMPLES OF AQUIFER DETECTION AND IDENTIFICATION

2.5.1 Atlantic margin, offshore United States

A body of low-salinity groundwater located off the coast of New Jersey (NJ) and Martha's Vineyard, Massachusetts (MA) (Figure 2.6) is believed to be part of a very large aquifer system extending offshore the United States Atlantic margin (Gustafson et al. 2019). This aquifer system is significant because it may be comparable in size to some of the largest known onshore aquifers in North America, such as the Ogallala aquifer.

In Figure 2.6, yellow circles represent wellbores with low pore fluid salinities (<15 ppm), while pink circles represent wellbores with high pore fluid salinities (>15 ppm). White triangles are passive magnetotelluric (MT) stations and white dashed lines are 2D controlled source electromagnetic (CSEM) profiles. The yellow grid of lines indicates where the electromagnetic (EM) data is interpreted to support an area with low-salinity pore water. The red line in the northeast (NE)—next to the 2D CSEM—indicates the location of the 2D seismic line in Figures 2.7 and 2.9.

2.5.1.1 Well data acquired in the 1970s and the 2010s

Prior to the 1970s, groundwater was known to exist onshore thanks to a large number of wells that had been drilled, thus establishing a local aquifer system. As more wells were drilled and local variations were found in the salinity of the pore fluid, the hypothesis emerged that several aquifer systems may be present, but their distribution and volumes were not well understood.

22 Integrated Aquifer Characterization and Modeling

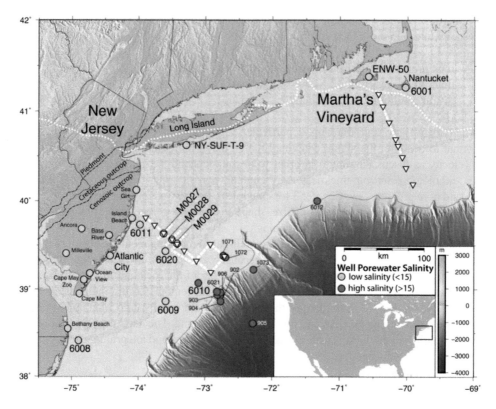

Figure 2.6[1] Location map of the aquifer off the US Atlantic continental shelf, NJ and Martha's Vineyard, MA. (Modified after Gustafson et al. 2019, refer to note 1 at the end of this chapter.)

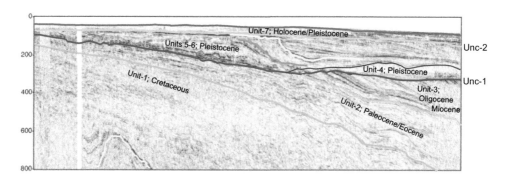

Figure 2.7 Seismic interpretation offshore the NE US continental shelf. (Adapted from Siegel et al. 2014.)

Aquifer presence and identification 23

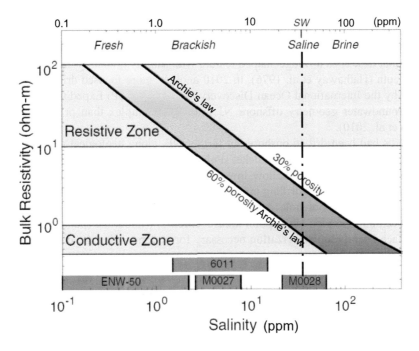

Figure 2.8[2] Expected range in total resistivity calculated as a function of water salinity, using Archie's law for clastic sediments with porosity varying from 30% to 60% encountered in nearshore wells. (Modified after Gustafson et al. 2019, refer to note 2 at the end of this chapter.)

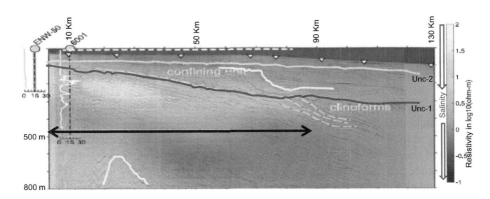

Figure 2.9[3] Bulk resistivity profile plotted using a logarithmic scale with colors similar to those in Figure 2.8 and superimposed onto the 2D seismic profile of Figure 2.7. (Modified after Gustafson et al. 2019, refer to note 3 at the end of this chapter.)

In the early 1970s, several oil and gas companies drilled nearshore the Atlantic margin in search of HC, but encountered fresh water in shallow sediments instead. The scientific drilling Atlantic Margin Coring Project (AMCOR) (1976) confirmed the existence of a low-salinity groundwater system as far as 120 km offshore NJ. Geological models were then developed that described a wedge-shaped feature with salinity gradually increasing seaward and with depth (Hathaway et al. 1976). In 2010 another, more focused drilling project was completed by the International Ocean Discovery Program (IODP) Expedition 313 showing that the groundwater geometry offshore NJ was more complex than previously thought (Mountain et al. 2010).

More wells had been drilled onshore but further NE, along geological trend at Martha's Vineyard (Hall et al. 1980) and Nantucket (Folger et al. 1978). These wells also supported the hypothesized presence of a shallow, low-salinity groundwater system possibly extending offshore, but no actual onsite drilling information was available to confirm this hypothesis.

In conclusion, while borehole data does entail direct sampling of subsurface aquifer systems, it can only provide 1D or point **measurements,** without the spatial correlation and the subsurface lateral **characterization** necessary for understanding the distribution of offshore groundwater systems.

2.5.1.2 New well and seismic data acquired by IODP-313 in the 2010s

The IODP (Integrated Ocean Drilling Program) 313 expedition acquired new well data and incorporated new 2D seismic data off Martha's Vineyard. The location and direction of the seismic (northwest, NW, to southeast, SE) is similar to that of the 2D CSEM profiles shown in the northern part of Figure 2.6 (the 2D CSEM is shown by a white line, while the 2D seismic is shown by a red line).

When integrating the seismic with the well control, Siegel et al. (2014) identified a system of Pleistocene age sediments with on-lap characteristics, and interpreted these young sediments as forming a top seal for the groundwater system encountered offshore (Figure 2.7). Seven stratigraphic units (labeled here U1 through U7) were identified from the Cretaceous to present day (Siegel et al. 2012). U1 (Cretaceous) and U2 (Paleocene and Eocene) consist of pelagic sedimentation of sands and carbonate muds. U3 resulted from an increase in siliciclastic from the Oligocene to the Miocene (Steckler et al. 1999) and consists of silts and clays. Regional unconformity U1 is believed to record the first Pleistocene shelf crossing glaciation offshore MA (Siegel et al. 2012). It likely forms a top seal to the underlying stratigraphy. U4 deposited glacigenic sediments in the late Pleistocene and is expected to consist of poorly sorted silts and clays. U5 and U6 consist of high siliciclastic input from high-amplitude sea level changes during the Pleistocene and likely contain sequences of sand, silt and clay. The units are capped by another unconformity U2 resulting from glacial outwash during the late Pleistocene and Holocene, and followed by recent siliciclastic sedimentation (U7). U2 marks the regional shallow sequence boundary that formed during the last sea level fall (Siegel et al. 2014).

The integrated interpretation of 2D seismic and wells provides a reasonable regional stratigraphic interpretation of the area. However, given that seismic is generally not sensitive to water salinity, it cannot provide information on whether several disconnected systems are present as opposed to a single groundwater system. Nonetheless, previously combined borehole and seismic data inferred either a system of multiple, localized water reservoirs within the mid-shelf offshore NJ (Lofi et al. 2013) or an aquifer system with limited offshore extent (Masterson et al. 2015).

2.5.1.3 Modern potential fields data acquired in the late 2010s

In 2015, geophysicists from Lamont-Doherty and the USGS acquired 2D EM and MT profiles using technology developed for petroleum exploration (Constable 2013). EM methods measure bulk electrical resistivity. Even though these methods are low frequency and hence low resolution, they are nevertheless effective at detecting lateral changes in the bulk resistivity of the subsurface.

The electrical resistivity of sediments is primarily a function of the porosity of the sediments and the resistivity of the formation fluid (Archie 1952). Archie's law is often written as follows:

$$S_w = \sqrt[n]{\frac{aR_w}{\emptyset^m R_t}} \qquad (2.3)$$

where
S_w = water saturation (fraction),
n = saturation exponent (usually equal to 2),
a = empirical constant (near unity),
R_w = formation water resistivity [ohm-meter (Ω-m)],
ϕ = porosity (fraction),
m = cementation exponent (usually equal to 2),
R_t = bulk resistivity (of uninvaded formation, in Ω-m).

The water salinity–resistivity relationship discussed previously (Figure 2.3, after Schlumberger 1998) was used for a fixed temperature estimated at around 10 °C to translate measured subsurface bulk resistivity (R_t) into aquifer salinity.

Any single particular shallow stratigraphic unit in Figure 2.6 exhibits a limited range in porosity, so we can expect a direct logarithmic correspondence between bulk resistivity and water salinity, as modeled in Figure 2.8. The top line assumes a unit with ~30% porosity while the bottom line assumes a unit with ~60% porosity. In other words, a key feature of the resistivity data lies in its ability to differentiate within a given stratum between more resistive low-salinity water and more conductive high-salinity water. In Figure 2.8, sediments saturated with sea water (defined as having 35 k ppm salt content) appear in light-to-dark blue colors, while sediments that are likely to house a lower-salinity aquifer (salinity less than 15 k ppm) are more resistive and shown in colors ranging from light green to orange and yellow.

Gustafson et al. (2019) describe in detail their geophysical quantification of the aquifer system. They combine EM/MT with the seismic reflection data (Figure 2.9) that was used as depth calibration for the stratigraphy, and they interpret a system of laterally continuous aquifers extending up to 90 km offshore, at which point the anomaly stops and seems to be bound laterally by a series of clinoforms (Figure 2.7). Figures 2.8 and 2.9 are plotted using the same color scale for resistivity and water salinity in the aquifer. Figure 2.9 shows a resistivity anomaly nearshore that begins at a depth of about 120 m below the seafloor and extends to a depth of approximately 380 m. The anomaly is interpreted as showing salinity increasing (and resistivity decreasing) gradually as one moves offshore. As in Figure 2.8, the color scale uses an arbitrary 15 ppm cut-off in fluid resistivity (yellow-green to blue transition), implying a low salinity (higher resistivity) anomaly extending out to approximately 90 km offshore.

We agree with Gustafson et al. (2019), whose interpretation implies a continuous and widespread distribution of low-salinity water across the shelf. This distribution is more regional in scale than previously thought. Preliminary calculations estimate the volume of the aquifer system at approximately 2,800 km³ of water with < 15 ppm salinity, assuming a dimension of approximately 350 km with an average thickness of 220 m and 45% porosity in a north-to-south (N–S) direction. The aquifer's size, if confirmed by future offshore drilling, is comparable to known giant onshore aquifers such as the Ogallala aquifer, estimated at 3,608 km³ (2,925 MM acre-feet) in 2005 by the USGS.

In conclusion, the work of Gustafson et al. (2019) unlocked a new understanding of the aquifer system offshore NJ and Martha's Vineyard, MA as being in hydrodynamic equilibrium. They interpreted fresh water recharge as coming from melting glacial deposits on the MA shelf in the west. The continental shelf was exposed during the last ice age, when the sea level was significantly lower than it is today. Rainfall and fresh water recharge from creeks and rivers flushed fresh water toward the east. Sea level rise subsequently inundated the area, depositing a blanket of fine sediments that served as a top seal trapping the fresh water. Saline water recharge comes from the deep underlying salt deposits located further east offshore. The entire system appears to be in hydrodynamic equilibrium, with sediments containing fresh water in the west and saline water originating from brines upwelling along faults in the east. This results in a pattern of gradually increasing salinity from onshore to offshore. As shown in Figure 2.10, a large body of low-salinity onshore water resulting from glacial melts

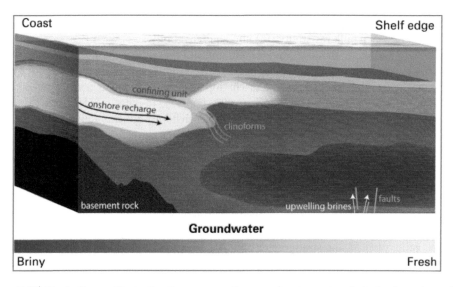

Figure 2.10[4] Block diagram illustrating the concept of a ground water system in hydrodynamic equilibrium offshore NJ. (Gustafson et al. 2019, refer to note 4 at the end of this chapter.)

feeds into an offshore high-salinity system originating from upwelling brines that passed through underlying salt deposits.

2.5.2 A seismic example: offshore Angola

When a HC reservoir is first discovered, best practices dictate that the potential extent of associated aquifers be identified and defined as soon as possible so as to establish the best strategy for recovering the resources. This early interpretation of the aquifer is then augmented during the field appraisal phase by teams tasked with devising possible recovery mechanisms and defining an optimum field development plan (FDP). Their analysis, which needs to be completed prior to the field investment decision (FID) that precedes production, drives the estimation of potential infrastructure and of capex requirements. The importance of establishing an early assessment of aquifers and their relationship to optimizing the recovery of resources will be discussed in detail in Chapter 7.

2.5.2.1 Descriptive seismic attributes

In the early appraisal stage that follows immediately after an exploration discovery, it usually suffices to identify the aquifer (if possible) and to define its areal extent in order to calculate its potential volume. In certain cases, seismic calibrated to a local discovery well can help identify the aquifer away from the well and estimate its extent fairly accurately, as shown in our next example from an oil field offshore Angola. In other cases where no well has been drilled nearby, it is still possible to use seismic in a self-calibrated manner (without well control) by interpreting relative changes in seismic character (amplitude, frequency, phase and velocity).

However, 3D seismic is best leveraged when tied to at least one existing well for local calibration on account of the relatively poor vertical resolution of seismic and of the necessity to calibrate rock properties in a given area to a known control point (the well). As a result, a typical more quantitative seismic-based aquifer identification workflow involves three key steps:

(1) Performing a well tie

The first step involves providing a detailed seismic synthetic well-tie (a zero-offset or acoustic synthetic tie is sufficient at this stage, but an elastic synthetic tie is better if pre-stack amplitude analysis is desired) to demonstrate an understanding of the main character of the seismic, as well as the properties—and limitations—of the recorded seismic wavelet. Indeed, the seismic wavelet recorded at a given depth is different from the wavelet generated at the surface due to the complexities of wave propagation and seismic attenuation in the Earth. The well tie provides a direct correlation between the stratigraphy encountered within a well and the seismic profiles. If the area is in "reconnaissance mode" and there are no wells available, the seismic interpreter should at least provide a detailed stratigraphic column for the location and describe how it ties to the regional structural-stratigraphic events chart for the overall basin.

(2) Mapping away from the well (or area of regional understanding)

The second step is the prerogative of the seismic interpreter, who defines key stratal packages and/or unconformities, and maps findings in the form of 3D seismic horizons. Each horizon is a 2D surface (which may be interrupted by faults or unconformities).

Seismic attributes are then extracted around said horizons. For example, one might extract a seismic attribute within a window around a particular horizon or in between a top horizon and a base horizon. The more accurately picked the horizons, the more interpretation control there is on the stratigraphic interval sampled by seismic. As discussed in Section 2.4.2.1, a large number of seismic attributes can be extracted, but the most common are related to either amplitudes (e.g. RMS amplitudes or sum of negative amplitudes, SNA), velocity, frequency or phase—all calculated within the sampled seismic interval.

Defining an appropriate seismic analog for comparison can be useful, though finding the "right" analog is not always straightforward. It is important that the seismic interpreter define the key assumptions of a geological model and describe how they translate into observed seismic signatures.

(3) Identifying fluid contact(s)

It is sometimes possible to identify fluid contacts using seismic (3D or 4D), although the seismic character of aquifers on stack seismic sections is usually quite opaque, making aquifers difficult to map. Nevertheless, pre-stack seismic techniques can provide significant help in identifying lithologies from seismic, even in the down-dip aquifer portion of a reservoir system, away from the oil. An example of seismically identified oil–water contact (OWC) will be discussed next.

2.5.2.2 Using seismic to identify and map the aquifer

In the following example (after Connolly et al. 2002), the HC accumulation is a three-way closure against a prominent fault that forms a NE edge to the trap. The oil trapped against the fault extends down-dip into a series of turbidite sands presenting channelized features in a N–S direction. Figure 2.11a) shows the trap and the OWC (shown in dashed yellow) as identified from the stacked seismic. In Figure 2.11b), pre-stack seismic data was processed to highlight and differentiate the oil-saturated reservoirs (in orange) from the water saturated sands (in light blue); this is sometimes called a seismic FI volume. Note that the aquifers are lower amplitude, but quite visible on this FI seismic volume. Conversely, in Figure 2.11c), pre-stack data was optimized to distinguish sands (in yellow and orange) from shales (in blue); this is sometimes called a seismic LI volume (see Section 2.4.2.2 on how FI and LI seismic volumes are created through χ angle rotations).

Deep-water mid-slope turbidites offshore Angola form extensive, highly sinuous channel and lobe systems with highly complex reservoir architectures (Kolla et al. 2001). The details of their geometry would be impossible to predict from wells alone, without insight from seismic (Doughty-Jones et al. 2017). Readers can grasp the complexity of the channel system in Figure 2.11. A main reservoir fairway measuring approximately 8 km in width and more than 10 km in length is pictured with details visible in three separate channel systems (channel belts) west to east. As the patchy patterns of the seismic lithology volume

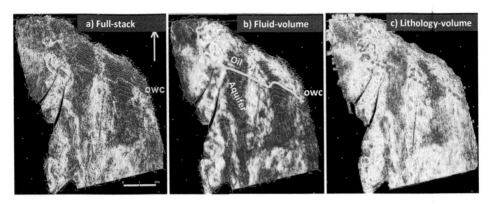

Figure 2.11 An example of fluid identification (oil versus water) and lithology detection (sands versus shales) using pre-stack seismic data from a deep-water Angola discovery. The seismic maps show average seismic amplitudes, extracted using band-limited impedance volumes derived from (a) the seismic-stack, (b) the seismic-fluid volume, and (c) the seismic-lithology volume. The OWC is traced on all three volumes for reference. (Modified after Connolly et al. 2002.)

of Figure 2.11c) show, connectivity between different channels and channel subsegments is likely quite tortuous.

All three maps are entirely data-driven (seismic driven), i.e. they were extracted from the data and their generation did not require any kind of imposed geological or engineering model.

Figure 2.12 shows a 3D perspective of the reservoir-aquifer system for the same area as Figure 2.11. Aquifer support is expected to be directional, coming from the down-dip extensions of the high-permeability channel sands. The channel sands are high quality, but their connectivity may not necessarily be well developed, as shown by the patterns in the aquifer portion.

In conclusion, seismic offers powerful techniques that can serve to identify the presence of aquifers and to calculate their volumes, but these techniques are insufficient to characterize aquifers in terms of their strength and effectiveness. To do that requires an integrated characterization of the aquifers, which is the subject of Chapters 4 and 5.

2.5.3 Detection of a hydrodynamic system

Describing the equilibrium state of fluids in the subsurface on a very large scale and grasping the full potential extent of aquifers requires a considerable amount of data as well as its multidisciplinary integration. For example, the presence of a tilted aquifer requires an understanding of the hydrodynamics of water flow within the aquifer as discussed below. Generally, the aquifer pressure decreases in the direction of OWC dip, whereas the oil gradient remains the same.

2.5.3.1 The Cretaceous Nahr Umr Lower Sands, offshore Qatar

Wells (1988) described the presence of a tilted aquifer around the giant North field located offshore Qatar, in the Persian Gulf. The tilted OWC is caused by hydrodynamic flow of high

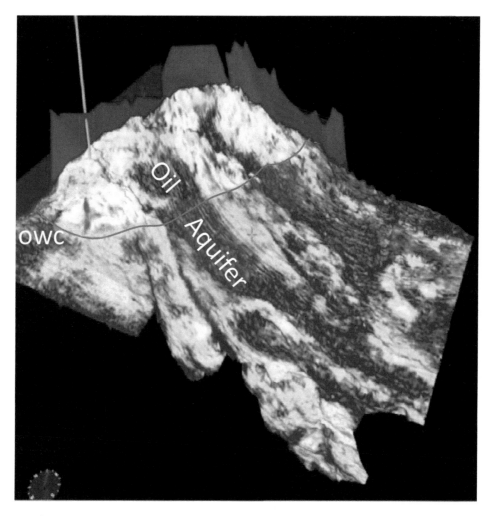

Figure 2.12 3D perspective of the turbidite channel sand oil-aquifer system for the Angola example in Figure 2.11. (Derived from Connolly 2010. SEG Spring Distinguished Lecture. SEG©2010, *reprinted by permission of the SEG whose permission is required for further use.*)

potential gradients across the field within the Nahr Umr Lower Sands. All measurements and observations—including log correlations, test results and structural maps—indicated that the distribution of oil and gas within Cretaceous reservoirs was not compatible with simple structural trapping. Wells' argument was controversial when first introduced, and it remained disputed until continued production from the field generated more evidence from dynamic data.

As discussed by Wells (1988), water flow in sedimentary basins is driven by geographic variations in water potential. During the early phase of basin development, water flow is directed upwards and up-dip toward the basin margins because of the relatively higher water potential in the basin center created by the expulsion of water during compaction and clay mineral diagenesis. With basin maturation, the water-potential system can change, depending on the formation of other topographic reliefs around it. For example, elevated water tables can create high water potentials at the basin margins (above mean sea level). This potential difference drives water flow inward and down-dip toward the topographic depressions in the basin center, creating the possibility of upward discharge.

In a tilted aquifer environment, water potential drives water toward the discharge zones in the low-potential area in the center of the Persian Gulf (Wells 1988). Wells (1988) mentioned that the compaction/dewatering of Cretaceous and Tertiary sediments associated with the development of the Persian Gulf geosyncline during the Tertiary may have contributed to the excess internal water potential on the Iranian margin of the Persian Gulf.

The inward flow of relatively fresh water through the aquifer could have a profound effect on HC distribution, reservoir rock diagenesis and oil chemistry. For example, due to the presence of this large tilted WOC, the measured pressure data (MDT) indicated three different WOCs in this region. Water was tested from what is believed to be a good-quality connected aquifer on the southeast flank of the Nahr Umr Lower Sands accumulation, 10 m (33 ft) above the deepest tested oil in a nearby well.

Wells et al (1988) mentioned that in addition to the recognized geometric aspects, lateral and vertical variations in formation-water-salinity are to be expected, as evidenced by low salinity (20 k to 40 k ppm sodium chloride, NaCl) in the southwest (SW) and high salinity (150 k ppm) in the NE. As water flows downward toward a regional potential sink located on the NE flank of the Nahr Umr oil accumulation, formation water becomes increasingly more saline due to dilution with interstitial water. Wells concludes that the region where hydrodynamic trapping may occur extends over a distance of almost 2,000 km, from Iraq to Oman. For additional discussion on tilted WOC, please see section 3.4.3 in Chapter 3.

2.6 AQUIFER EXTENT SUMMARY AND CONCLUSIONS

It is important to identify the extent of aquifers from a combination of well information (direct observations) and remote sensing. Readers should keep in mind that aquifers can have gigantic scales, as is the case for all three examples discussed in this chapter.

- For the example off the NE coast of the US, water movement is interpreted as occurring through a system of lower salinity (<15 k ppm) that appears connected at a scale of approximately 80 km in an east–west (E–W) direction and approximately 350 km in a N–S direction.
- For the Angola example, each of the three individual channel belts identified measures approximately 10 km in length by 1–2 km in width. A visual inspection of Figures 2.11 and 2.12 suggests an expected aquifer/oil ratio of < 5:1 for the westernmost channel, < 3:1 for the central channel, and < 2:1 for the eastern channel.
- Geology knows no borders for the Nahr Umr Lower Sands Cretaceous aquifer. It took experts a long time to fully integrate the well pressures and production data for the area, and to grasp the huge extent (nearly 2,000 km) of the aquifer that is possibly shared by the countries spanning the region (Oman to Iraq).

Exploration and appraisal in regions that are likely impacted by hydrodynamic conditions should include an uncertainty analysis due to possible ranges in the distribution of fluids in the subsurface.

An Example of Volumetric Uncertainty Estimation: the US Atlantic Margin Aquifer System

- It is unclear from the extrapolated 2D seismic data whether the aquifer is truly connected in a N–S sense over a range of up to 350 km. If the aquifer were occurring only around the immediate NJ area, it could be connected over less than 200 km; conversely, it could extend over 450 km north beyond Martha's Vineyard.
- Similarly, when considering the E–W direction, the easternmost part of the aquifer may extend < 60 km offshore because, as seen in Figure 2.9, the resistivity anomaly persists beyond 60 km but becomes significantly weaker; the aquifer could otherwise extend further offshore where the anomaly is weak for a distance of perhaps 90 km or more.
- Other sources of uncertainty exist when calculating the range of volume for the aquifer, including its average thickness, average porosity and the limit where salinity drops below 15 ppm. This last uncertainty on salinity limits is already captured under the uncertainty for the E–W extent of the aquifer. If we assume the input ranges shown on the left-hand side of Figure 2.13, the output is a statistical distribution of possible aquifer volumes shown on the right-hand side of Figure 2.13. The figure indicates that the size of the aquifer has a 90% chance of being 1,016 km^3, a 50% chance (most likely) of being 2,317 km^3 and a 10% chance of being 4,433 km^3.

The volumetric uncertainty regarding aquifer size is significant, with a P10/P90 ratio of almost 5, and it will remain so until more data or information becomes available. By providing a range of input parameters that is slightly skewed to the left (meaning that the distribution of input values for the N–S extent, W–E extent and for thickness are more likely to be smaller than the deterministic case), the output shows a distribution of aquifer size that is slightly asymmetric with a most likely P50 value of around 2,317 km^3.

An aquifer's overall effectiveness is expected to be impacted not only by its total volume, but also by its connectivity and strength. The latter properties are discussed in Chapter 7.

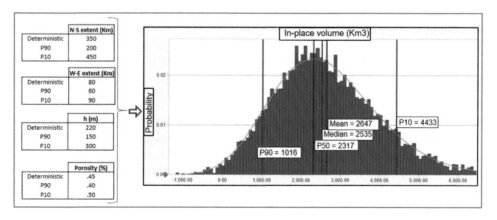

Figure 2.13 Volumetric uncertainty calculation for the offshore NE US Atlantic margin aquifer.

NOTES

1 This work is licensed under the Creative Commons Attribution 4.0 International License. To view a copy of this license, visit http://creativecommons.org/licenses/by/4.0 or send a letter to Creative Commons, PO Box 1866, Mountain View, CA 94042, USA.
2 This work is licensed under the Creative Commons Attribution 4.0 International License. To view a copy of this license, visit http://creativecommons.org/licenses/by/4.0 or send a letter to Creative Commons, PO Box 1866, Mountain View, CA 94042, USA.
3 This work is licensed under the Creative Commons Attribution 4.0 International License. To view a copy of this license, visit http://creativecommons.org/licenses/by/4.0 or send a letter to Creative Commons, PO Box 1866, Mountain View, CA 94042, USA.
4 This work is licensed under the Creative Commons Attribution 4.0 International License. To view a copy of this license, visit http://creativecommons.org/licenses/by/4.0 or send a letter to Creative Commons, PO Box 1866, Mountain View, CA 94042, USA.

REFERENCES

Archie, G.E. (1952). Classification of carbonate reservoir rocks and petrophysical considerations. *AAPG Bulletin* 36 (2): 218–298.

Castagna, J.P., Swan, H.W., and Foster, D.J. (1998). Framework for AVO gradient and intercept interpretation. *GEOPHYSICS* 63 (3): 948–956. http://dx.doi.org/10.1190/1.1444406.

Chopra, S. and Marfurt, K. (2005). Seismic attributes—A historical perspective. *GEOPHYSICS* 70 (5). https://doi.org/10.1190/1.2098670.

Connolly, P. (2010). Robust workflows for seismic reservoir characterization. SEG Spring Distinguished Lecture. *Recorder: Official publication of the Canadian Society of Exploration Geophysicists* 35 (4): 7–10. https://seg.org/Education/SEG-on-Demand/id/6209/distinguished-lecture-recordings-robust-workflows-for-seismic-reservoir-characterization (accessed March 11, 2022).

Connolly, P., Schurter, G., Davenport, M. and Smith, S. (2002). Estimation net pay for deep water turbidite channels offshore Angola. EAEG 64th Conference and Exhibition, extended Abstract G-28. Available from https://doi.org/10.3997/2214-4609-pdb.5.G028.

Connolly, P. A. (1999). Elastic impedance. *The Leading Edge* 18 (4): 438–452. Available from http://dx.doi.org/10.1190/1.1438307.

Constable, S. (2013). Review paper: Instrumentation for marine magnetotelluric and controlled source electromagnetic sounding. *Geophysical Prospecting* 61: 505–532.

Doughty-Jones, G., Mayall, M., and Lonergan, L. (2017). Stratigraphy, facies and evolution of deep-water lobe complexes within a salt-controlled intraslope minibasin. *AAPG Bulletin* 101 (11): 1879–1904. https://doi.org/10.1306/02071716056.

Dutta, N.C., Bachrach, R. and Mukerji, T. (2021). *Quantitative Analysis of Geopressure for Geoscientistsand Engineers*. Cambridge: Cambridge University Press.

Eppelbaum, L.V. (2019). *Geophysical Potential Fields: Geological and Environmental Applications*. Computational Geophysics 2. Amsterdam: Elsevier. https://doi.org/10.1016/B978-0-12-811685-2.12001-6.

Folger, D.W., Hathaway, J.C., Christopher, R.A., Valentine, P.C. and Poag, C.W. (1978). *Stratigraphic Test Well, Nantucket Island, Massachusetts*. US Geological Survey Circular 773. Washington, DC: Department of the Interior. Available from https://pubs.usgs.gov/circ/1978/0773/report.pdf.

Flemings, P. (2021). *A Concise Guide to Geopressure: Origin, Prediction, and Applications*. Cambridge: Cambridge University Press.

Freeze, R.A. and Cherry, J.A. (1979). *Groundwater*. Englewood Cliffs, NJ: Prentice-Hall.

Gustafson, C., Key, K. and Evans, R.L. (2019). Aquifer systems extending far offshore on the U.S. Atlantic margin. *Scientific Reports* 9: article 8709). https://doi.org/10.1038/s41598-019-44611-7.

Hall, R.E., Poppe, L.J. and Ferrebee, W.M. (1980). *A Stratigraphic Test Well, Martha's Vineyard, Massachusetts*. Washington, DC: US Geological Survey.

Han, D.-H. and Batzle, M. (2014). FLAG fluid calculator. Paper presented at the joint University of Houston/CSM Fluids/DHI Consortium.

Hathaway, J.C. et al. (1976). *Preliminary Summary of the 1976 Atlantic Margin Coring Project of the US Geological Survey*, 76–844. US Geological Survey Open-File Report.

Kolla, V., Bourge, Ph., Urruty, J.-M. and Safa, P. (2001). Evolution of deep-water Tertiary sinuous channels offshore Angola and implications for reservoir architecture. *AAPG Bulletin* 85: 1373–1405. https://doi.org/10.1306/8626CAC3-173B-11D7-8645000102C1865D.

Konikow, L.F., and Bredehoeft, J.D. (2020). *Groundwater Resource Development: Effects and Sustainability*. Guelph, Ontario: The Groundwater Project. https://gw-project.org/books/groundwater-resource-development.

Lofi, J. et al. (2013). Fresh-water and salt-water distribution in passive margin sediments: Insights from Integrated Ocean Drilling Program Expedition 313 on the New Jersey Margin. *Geosphere* 9: 1009–1024.

Masterson, J.P. et al. (2015). *Hydrogeology and Hydrologic Conditions of the Northern Atlantic Coastal Plain Aquifer System from Long Island, New York, to North Carolina*. Reston, VA: US Geological Survey.

Mountain, G.S. et al. (2010). *Proceedings of the Integrated Ocean Drilling Program*, vol. 313. Tokyo: Integrated Ocean Drilling Program Management International. Available from https://doi.org/10.2204/iodp.proc.313.101.2010.

Pederson, K.S., Christensen, P.L and Shaikh, J.A. (2014). *Phase Behavior of Petroleum Reservoir Fluids*. 2nd edn. Boca Raton, FL: CRC Press.

Peltenburg, E. (2012). East Mediterranean water wells of the 9th–7th millennium BC. In: F. Klimscha (ed.), Wasserwirtschaftliche Innovationen im archäologischen Kontext: Von den prähistorischen Anfängen bis zu den Metropolen der Antike, 69–82. Rahden/Westphalia: Leidorf.

Prensky, S. (1992). Temperature measurements in boreholes: An overview of engineering and scientific applications. *The Log Analyst* 33: 313–333.

Rutherford, S.R., and Williams, R.H. (1989). Amplitude-versus-offset variations in gas sands. *GEOPHYSICS* 54: 680–688. Available from http://dx.doi.org/10.1190/1.1442696.

Schlumberger (2013). Log interpretation charts (webpage). www.slb.com/-/media/files/premium-content/book/log-charts/chartbook.ashx.

Siegel, J., Dugan, B., Lizarralde, D., Person, M., DeFoor W. and Miller, N. (2012). Geophysical evidence of a late Pleistocene glaciation and paleo-ice stream on the Atlantic Continental Shelf offshore Massachusetts, USA. *Mar. Geol.* 303–306: 63–74.

Siegel, J., Person, M., Dugan, B., Cohen, D, Lizarralde, D. and Gable, C. (2014). Influence of late Pleistocene glaciations on the hydrogeology of the continental shelf offshore Massachusetts, USA. Geochemistry, Geophysics, Geosystems 15 (12): 4651–4670. https://doi.org/10.1002/2014GC005569.

Steckler, M.S., Mountain, G.S., Miller, G.M. and Christie-Blick, N. (1999). Reconstruction of Tertiary progradation and clinoform development on the New Jersey passive margin by 2-D back-stripping. *Mar. Geol.* 154, 399–420.

Summerlin, P. (2011). Water salinity diagram. CC BY-SA 3.0. https://commons.wikimedia.org/w/index.php?curid=13274737.

Taner, M.T., Koehler, F. and Sheriff, R.E. (1979). Complex seismic trace analysis. *GEOPHYSICS* 44: 1041–1063. Available from https://doi.org/10.1190/1.1440994.

Theis, C.V. (1940). The source of water derived from wells—Essential factors controlling the response of an aquifer to development. *Civil Engineering* 10: 277–280. Available from www.eqb.state.mn.us/sites/default/files/documents/Source_of_Water_Derived_from_Wells.pdf.

Wells, R.A. (1988). Hydrodynamic trapping in the Cretaceous Nahr Umr Lower Sand of the North Area, offshore Qatar. *Journal of Petroleum Technology* 40 (3): 357–361. https://doi.org/10.2118/15683-PA.

Whitcombe, D.N., Connolly, P.A., Reagan, R.L. and Redshaw, T.C. (2000). Extended elastic impedance for fluid and lithology prediction. *GEOPHYSICS* 67 (1): 63–67. Available from https://doi.org/10.1190/1.1451337.

Widess, M.B. (1973). How thin is a thin bed? *GEOPHYSICS* 38 (6): 1176–1180. http://dx.doi.org/10.1190/1.1440403.

Zoeppritz, K. (1919). VIIb. Über Reflexion und Durchgang seismischer Wellen durch Unstetigkeitsflächen (On reflection and transmission of seismic waves by surfaces of discontinuity]. *Nachrichten von der Königlichen Gesellschaft der Wissenschaften zu Göttingen, Mathematisch-physikalische Klasse* 1919: 66–84. www.digizeitschriften.de/dms/img/?PID=GDZPPN002505290.

Chapter 3

Aquifer description and characterization

INTRODUCTION

This chapter discusses various aspects of aquifer description and characterization through the interpretation of data generated by three key subsurface disciplines at various scales of investigation, namely:

- geology
- geophysics
- petrophysics

Rather than compile an exhaustive review of the workflows employed by all three disciplines, this chapter aims instead to present the essential concepts that form the technical basis for the construction of realistic earth models, a topic that is more fully addressed in Chapters 4 and 5.

3.1 AQUIFER CHARACTERIZATION USING GEOLOGICAL INFORMATION

3.1.1 Regional geology: a play-based approach

Geologists account for the makeup and interaction of various geological bodies by means of their understanding of the genesis of geological entities in common past events—depositional or structural—within defined regions.

Establishing the characteristics of an area's regional geology entails retracing the context in which sediments are originally deposited and how they evolve through time. The historical unfolding of interrelated geological processes such as subsidence, accommodation and potential subsequent sediment deformation provides an important basis from which to understand the likely extent of the geological systems present at the time of deposition.

3.1.1.1 Play-based concepts

Geoscientists from private and/or public enterprises (e.g. Shell's play-based exploration, 2021) as well as government organizations (e.g. Miller 1982) have routinely applied exploration play analysis techniques to assess the potential for locating petroleum resources in various basins. The play-based workflow uses concepts spanning the geological scale, the main "building unit" being the *segment*. Geoscientists define basins, plays, sub-plays and

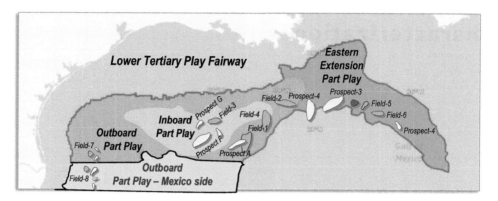

Figure 3.1 The Lower Tertiary slope play in the deep-water Gulf of Mexico basin.

segments that form prospects (undiscovered resources) and fields (discovered resources), as illustrated in Figure 3.1. As geology knows no borders, the outboard part play extends south into Mexico.

Common terms used in the petroleum industry are defined below.

- ***Basin***

A geological ***basin*** can take on different meanings depending on its location and containment.

Structural geologists define basins as containers created by tectonic processes such as rifting, thrusting, etc. In this context, the term basin is often used to define a geographic province, e.g. the North Sea. Thus, basins are usually tectonic areas formed by plate tectonic processes.

Sedimentary basins are depressions in the earth's crust filled with sediments in the geologic past, making them repositories of genetic information. Depressions may have been formed by plate tectonic forces initially, with sediments subsequently accumulating in these depressions. Continued deposition may have caused further depression or subsidence and thickening of the depositional sequences into growth wedges. Sedimentary basins tend to have very large scales on the order of tens to hundreds of kilometers in length and width, and thousands of meters in depth.

Groundwater basins are used for surface aquifers, and ***drainage basins*** define a river's system and catchment area.

- ***Play***

A ***play*** is a regionally extensive area with similar controls on hydrocarbon (HC) accumulation throughout its span; it is often tied to a defined stratigraphic interval. A play often encompasses a single petroleum system within a basin and consists of several part plays. In other words, a play is a collection of part plays that share common geologic elements and belong to a single HC system (i.e. common source and reservoir). Examples are the Miocene play and the Lower Tertiary play in the Gulf of Mexico (GoM).

- *Part Play*

A *part play* defines one or more segments that share a common set of play risks; in other words, segments in a part play are distinguished only based on segment-specific risk. The Western trend of the Paleocene play in the GoM is an example of a part play.

- *Segment*

As defined in Chapter 2, a segment is a proposed or known accumulation that can be characterized by a single distribution for the connected volume and a single probability of occurrence, P_g (or probability of geological occurrence); $P_g = 1$ for a known accumulation, and $P_g < 1$ for an undrilled accumulation.

- *Volume aggregation*

For O&G business decision purposes, volumes are often aggregated through fields and/or prospects within a sub-play, so that they share a common facility or export route. Prospects or fields and their extensions are no longer simple, large structural traps (anticlines) with a single given reservoir, as simple geological features of that type have almost all been identified and drilled by now. Instead, volume aggregation typically involves multiple segments that are integrated into a common development plan and combined for the purpose of optimizing economics (Figure 3.2).

As in the HC industry, selecting a good site for various uses of an aquifer system begins with a regional understanding of the geology at the basin scale and encompasses all of the elements defined above.

When analyzing a system of aquifers, volumes will be aggregated within a basin and will encompass selected areas (which may or may not incorporate depleted HC fields) as well as localized sites and/or compartments, so that water segments share a common transport route and/or facility. The subsurface "building block" is, once again, the segment. This aggregation or grouping of segments tied into a common injection/production site (Figure 3.3) enables taking the most effective business decision with respect to managing water resources.

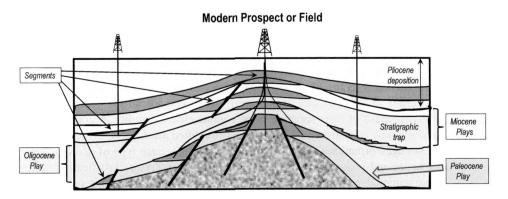

Figure 3.2 Aggregation of volumes around a modern petroleum field or prospect.

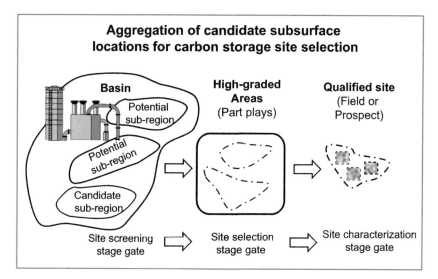

Figure 3.3 Aggregation of volumes around a system of aquifers. Note the parallel with Figure 3.1, from basins, to high-graded sub-areas and finally, local sites.

Because the workflows employed toward characterizing and qualifying potential carbon storage sites are complex and involve progressively increasing investments, they are broken into rigorous stage gates, in a similar manner to Oil & Gas projects.

3.1.2 Depositional systems and stratigraphy

The recognition of depositional systems is relevant to the characterization of aquifers, because depositional systems are a fundamental building block of the rock strata, regardless of whether a petroleum system is present for any basin.

Sequence stratigraphy, a practice developed by the O&G industry, incorporates core, outcrop, well and seismic information to interpret rock relationships within a chronostratigraphic framework of genetically related strata bounded by erosion or deposition surfaces or by their "correlative conformities." Sequence stratigraphy builds off sedimentary facies, a stratigraphic unit characterized by distinct lithological and/or age characteristics that usually reflects a common depositional origin. A group of sedimentary facies that are genetically linked by common processes and environments form a ***depositional system***.

While there are many types of depositional systems, most geologic facies that have yielded great results in O&G or that are currently being considered for geologic storage are sedimentary, because of their superior storage/volumetric capability. A good overview of depositional systems is summarized in Table 3.1 (modified after NETL's "Best Practices for Geologic Storage Formation Classification" (NETL 2010). Each type of geologic facies has different scales and properties. The geologic formation/reservoir classification system includes unconventional reservoirs (e.g. coalbeds) and igneous formations (e.g. stacked basalts). Note that shales are not included in the classification. Shales are composed of very fine clay and silt particles packed so closely together that they do not allow fluids to move

Table 3.1 Classification of the main depositional systems in the petroleum industry

Rock classification lithology	Geoscience Institute for O&G Recovery research classification in 1991			DOE's oil reservoir classification from 1990	Depositional system
Sedimentary	Clastic reservoirs	Delta	Delta/Fluvial dominated	Class 1 reservoirs	Deltaic
			Delta/Wave dominated		
			Delta/Tidal dominated		Coal/Shale
			Delta/Undifferenti-tated		
		Fluvial	Fluvial/Braided	Class 5 reservoirs	Fluvial
			Fluvial/Meandering		
			Fluvial/Undifferentitated		
		Alluvial fan			Alluvial
		Strand Plain	Barrier & Shoreface	Class 4 reservoirs	Strandplain
			back Barrier		
			Undifferentiated		
		Turbidites	Slope	Class 3 reservoirs	Turbidites
			Basin/Channels		
			Basin/Fans		
		Eolian	Clastics and/or Carbonates		Eolian
		Lacustrine	Clastics, carbonates, evaporites		Lacustrine
		Shelf			Shelf
	Carbonate reservoirs (>50% but can contain terrigenous sediments, quartz, felsdpar and evaporites)	Peritidal	Dolomites		
			Massive dissolution		
			Other		
		Shallow Shelf /Open	Dolomites	Class 2 reservoirs	Shallow shelf
			Massive dissolution		
			Other		
		Shallow Shelf /Open	Dolomites		
			Massive dissolution		
			Other		
		Reef	Dolomites		Shallow shelf
			Massive dissolution		
			Other		
		Shelf margin	Dolomites		
			Massive dissolution		
			Other		
		Slope/Basin	Other		

(continued)

Table 3.1 Cont.

Rock classification lithology	Geoscience Institute for O&G Recovery research classification in 1991	DOE's oil reservoir classification from 1990	Depositional system
Igneous			Basaltic flows
	Basalts		Interflow zones
	Granites		
Metamorphic			

Source: Modified after NETL (2010)'s "Best Practices for Geologic Storage Formation Classification."

through the matrix. Instead, fluid flow occurs in fractures—either natural or induced—and vertical flow is negligible when compared to horizontal flow along bedding planes. For the purpose of storage capability, shales are considered caprocks or seals. It is possible that future research will enable the use of fractured organic-rich shales as reservoirs for the geologic storage of liquids (e.g. Bakken shale, Marcellus shale), but they are currently considered not viable.

Important tools used for describing regional stratigraphy include:

- regional stratigraphic charts, which set the context for regional depositional events (Figure 3.4, left-hand side)
- tectono-stratigraphic event charts, which set the the context for regional structural events (Figure 3.4, right-hand side)
- regional cross-sections (Figure 3.5), which can made to emphasize either the structure (when plotted in depth) or the stratigraphy (when plotted relative to a particular stratigraphic interval)
- geological gross depositional environment maps or GDEs (Figure 3.6)

GDE maps are a mixture of paleogeography, paleobathymetry and lithology that indicate the distribution of reservoir, source or seal depositional facies for a particular stratigraphic interval.

In Figure 3.5, note both the significant vertical heterogeneity within individual wells and the lateral variability within zones between the wells. The reservoir section is a shoreface sequence, which is interpreted as having been incised by a valley system, normal to the underlying shoreline, and filled with coarse-grained fluvial/braided sediments.

As discussed in Chapter 2, aquifers can be sizeable and quite well connected over large areas. A striking example from the Bass Strait, depicted in Figure 3.7, involves a large number of gas fields producing from the L100-400 series sands, that display different pressure gradients in the HC leg while apparently sharing a common region aquifer. Furthermore, this aquifer appears to deplete fairly uniformly, as evidenced by the observations from more recent field developments that indicate the presence of the same regional aquifer with approximately 25 psi of depletion. Similar phenomena involving multiple fields that share a common aquifer

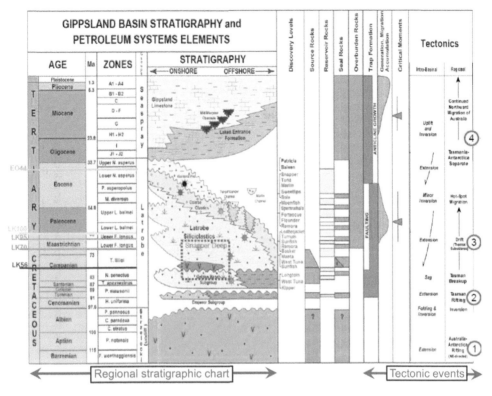

Figure 3.4 A regional stratigraphic chart (left-hand side) and the main tectono-stratigraphic events (right-hand side) for the Gippsland basin, offshore Australia.

at the basin scale have also been observed in carbonates, for example in the Campeche area offshore Mexico or offshore Qatar.

3.1.3 Structural model and faulting

In the petroleum industry, structural geology is essential for an integrated understanding of the subsurface. This structural understanding, which encompasses the genesis and preservation of HC traps, occurs at a wide range of scales, from the basin scale all the way down to the micro scale.

The structural context (pre-deposition) often provides the framework for the sedimentation history as well as the reason for the occurrence of igneous and metamorphic rocks in a basin. The structural evolution of an area (post-deposition) and the associated local rock deformation help further delineate the extent of potential traps and the likelihood for compartmentalization. Recent technological developments and workflows at the macro (tectonic), local (structural/faulting) and micro (fracture) scales have shed new light on the types of data that need to be obtained and have spurred the development of advanced algorithms or workflows used to describe subsurface deformation throughout geological time.

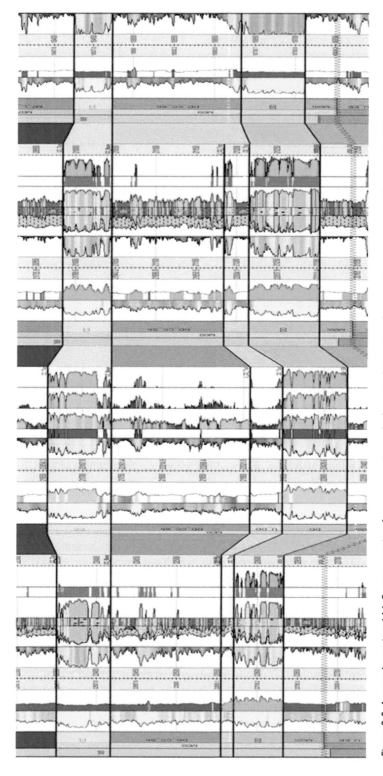

Figure 3.5 A pseudo-regional N–S cross-section from reservoir sands deposited in a shoreface environment over a distance of ~25 km.

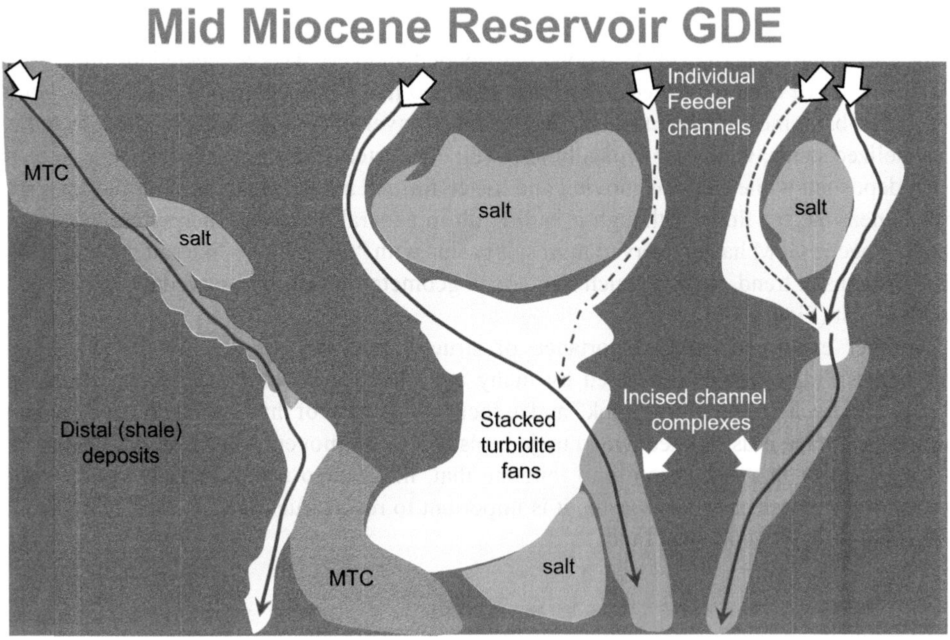

Mid Miocene Reservoir GDE

Figure 3.6 A gross depositional environment (GDE) map for the lower GoM Miocene play, derived from existing well control and seismic in the area.

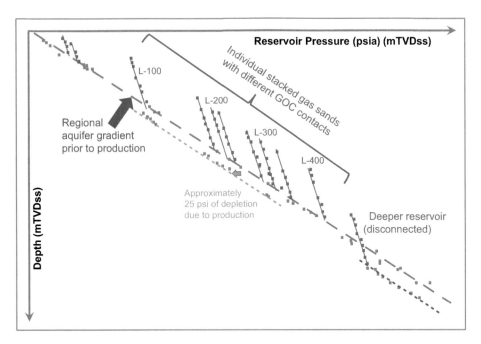

Figure 3.7 Schematic of the reservoir pressures from various fields in the Bass Strait, showing a very large aquifer that is shared at the basin scale.

3.1.3.1 Large-scale structural deformation

Structural evolution often plays a key role in the deposition of sands and their connectivity. For example, in the GoM, regionally extended and very well connected turbidite sheet sands may have been deposited before deep salt had had a chance to move much, i.e. while the paleo-bathymetry remained relatively flat. Conversely, more locally ponded and/or channelized sands, which are typically less well connected on a regional scale, tend to have been deposited while salt was moving and hence formed a relief defining multiple mini salt basins near the sea floor. Although it is difficult to generalize, many interpreters looking at the Paleogene GoM have observed mini salt basins with "egg carton"-like salt geometries in the inboard salt trend, while a much more open geometry has been observed in the outboard trend of salt (Figure 3.8).

Another example of the importance of structural movement is to be found in the "halokinesis" geometries described by many authors (Rowan et al. 1999) and shown in Figure 3.9. Each sedimentary "hook" and associated package of thinning sediment occurring near the salt face is associated with a major episode of salt movement and/or withdrawal.

The point of this discussion is to reiterate that, in order to understand the connectivity potential of deposits at a large scale, it is important to reconstruct these bodies' genesis and evolution through geological time.

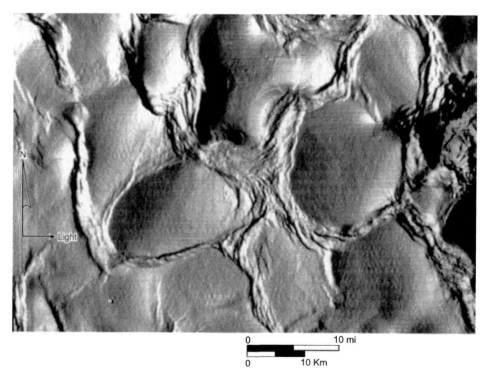

Figure 3.8 Example of GoM mini salt basins displaying "egg carton" geometry (Rowan et al. 1999; AAPG©1999, *reprinted by permission of the AAPG whose permission is required for further use*).

Aquifer description and characterization 47

Halokinesis sequence around a salt diapir

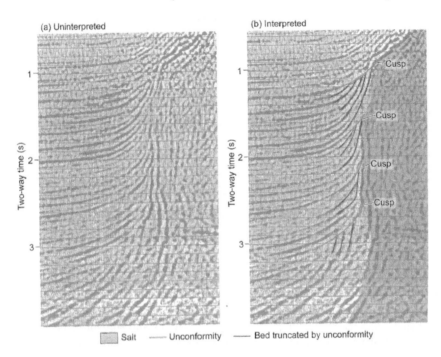

Figure 3.9 Example of halokinesis geometry observed around salt (Hudec and Jackson 1999, AAPG Memoir 99. AAPG©2011, *reprinted by permission of the AAPG whose permission is required for further use*).

3.1.3.2 Local scale structural features and fault traps

Local structural features are often related to faulting. The structural understanding of faults is paramount to assessing their potential to act as seals (barriers), baffles or conduits for fluid movement. An understanding of the stresses around faults—gained, for instance, by assessing fault orientation and dip with respect to regional stresses and by determining whether the faults are critically stressed, i.e. near failure—can yield great insights for fault seal analysis.

If no information is available to characterize a particular fault, the common practice is to assume that it will act as a barrier to flow. However, if there is sufficient information to describe the fault, it is possible to assess its likelihood of compartmentalizing a reservoir through a fault seal analysis workflow. Fault seal analysis, in its simplest form, involves analysis of the juxtaposition across a fault. As a rule of thumb, geologists assume that if (a) good quality sand is directly juxtaposed against another high quality sand and (b) the fault zone is not damaged (in the sense of *cataclasis* or grain crushing), then the fault is likely to allow the flow of fluids from one side to the other.

More complex fault geometries have been analyzed for leakage (Bretan 2017) based on the assumption that HC leakage through a water-wet fault zone occurs when the excess

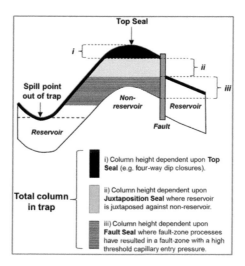

Figure 3.10 Three possible column height scenarios based on the characteristics of the fault shown in green (modified after Bretan 2017).

(buoyancy) pressure generated by the HC column exceeds the capillary threshold pressure of the fault zone material. The capillary entry pressure of the fault zone material can be approximated using the shale gouge ratio (SGR) related to the amount of clay "smeared" across the fault plane during fault displacement. Bretan (2017) defines three possible leak points across the fault (Figure 3.10) with corresponding potential HC heights for each of the following scenarios: (i) the fault does not seal; (ii) the fault juxtaposes sand (within the anticline) against non-permeable rock; and (iii) the fault has an added column due to the shale gouge ratio or to cataclasis.

Although Bretan's workflow (2017) applies to a HC trap, a similar workflow can be applied to aquifers, using aquifer pressures on each side of the fault. In this manner, any fault-bounded segment (whether or not it contains HCs) can be assessed as to its overall connectivity potential, i.e. whether it is a separate compartment or is connected to adjacent segments.

The degree of compartmentalization can be ranked or "graded" according to the effectiveness of the baffles/barriers between adjacent areas. In addition to being dependent on the properties of faults acting as segment boundaries, this baffle/barrier effectiveness is also a function of the permeability of the material/the amount of fluid that can flow through it. In turn, permeability is dependent on the quality of the reservoir and the fluids present.

3.1.4 Reservoir quality assessment

In Section 3.1.1, we showed the importance of having geological insight into how and when depositional systems form and their subsequent directional patterns. Geological facies determine how fluids will flow through porous media, so geologists and petrophysicists use facies and petrofacies as the basis for classifying rocks into potential flow units. They use geological analogs and modern outcrops, and they evaluate cores and well logs to assess which key facies are likely to exert a strong influence on fluid flow.

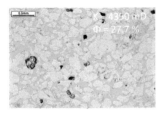  

Examples of thin sections displaying diminishing reservoir quality from left to right

Figure 3.11 An example of thin section analysis from two different samples.

Reservoir quality is typically assessed at the local scale (i.e. at the field or prospect scale). Key tools used to assess reservoir quality include log interpretation and core analyses.

3.1.4.1 Thin sections

Thin sections are used to assess mineralogy and grain size and shape, as well as to provide an indication of porosity. Figure 3.11, shows three different samples, with pore space appearing filled by blue epoxy. A higher quality reservoir is shown on the left while a lower quality reservoir appears on the right. In general, thin sections can be made from core or from drill cuttings and/or borehole cavings.

3.1.4.2 Core analyses

Sidewall cores (SWC) are obtained through percussion against a wellbore wall or through a rotary mini-drill and are often used as formation samples. The advantage of SWCs is the speed at which they are recovered for a low cost. Rotary sidewall cores (RSWC) are small 3D rock samples used for quantitative analyses. However, they are not considered to be of sufficient quality for relative permeability measurements and special core analysis. Instead, SWCs are generally used for calibrating well log data.

Full or whole cores (Figure 3.12) have the advantage of enabling the measurement of properties on a scale that is closer to that of the reservoir. Typical whole core sections can measure 9–36 m (30–120 ft) in length and are extracted using a core barrel made of steel. Whole cores are particularly important for heterogeneous formations, as they help evaluate the amount of heterogeneity within the reservoir (in the direction of the wellbore).

By means of routine core analyses (RCAL), whole cores allow for the accurate determination of porosity, single-phase permeability, fluid content, geologic age, as well as for estimating the local reservoir's likely productivity. Whole cores also enable special core analyses (SCAL), which include measurements of multi-phase flow properties and the determination of relative permeability, wettability, capillary pressure and electrical properties. Because SCAL is very time-consuming (typical project duration is several months and up to a year), they are also costly. To maximize sample quality, RCAL results are used to select samples that will undergo SCAL.

We mentioned that geologists and petrophysicists classify rocks according to facies. This is particularly important because facies form key building blocks for the construction of flow

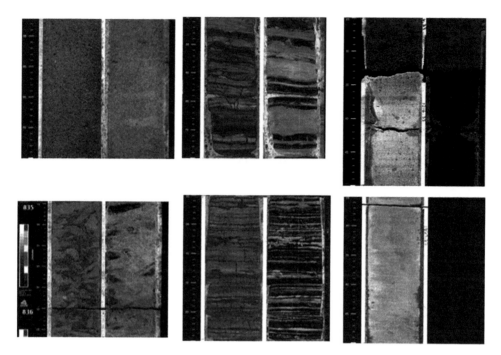

Figure 3.12 An example of a whole core after it has been slabbed and before it is sampled for analysis.

units. The process of facies classification starts with a detailed core description and analysis and is followed by and integrated with core measurements and log analysis. Typical generated plots include porosity-permeability trends by rock type (facies), an example of which is shown in Figure 3.13. The data points are plotted in color by facies (up to 12 of them in total) and subdivided by grain size. Kozeny-Carman permeability predictions (solid lines)—based on porosity, grain size and tortuosity—match the data fairly well for reasonable ranges in porosity.

3.1.5 Production geology

Production geology focuses on the understanding of flow units and flow barriers typically encountered within a reservoir. Production geology incorporates descriptions of the reservoir rocks, the HC fluids and the aquifer system. As discussed in Chapter 2, due to their usually large volumetric extent, aquifers often provide energy to the overall geological "plumbing" in the subsurface. The inputs to establishing a production geology framework are discussed in the remaining sections of this chapter.

Important issues associated with production geology include:

- An understanding of vertical displacement efficiency (including layering in the z direction) through the establishment of layers, baffles and faults.
- An understanding of rock quality, including facies, grain size, pore throats and diagenesis. Reservoir modifications beyond natural compaction and lithification often occur due to the circulation of fluids, which leads to diagenesis (more information on diagenesis is included in Chapter 4).

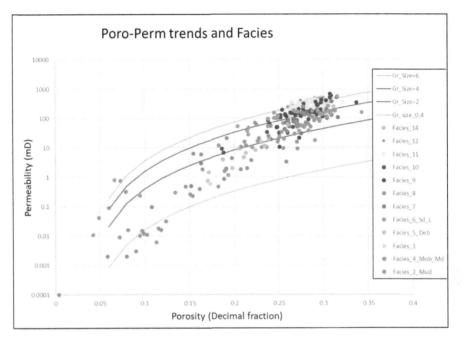

Figure 3.13 Facies subclassification corresponding to distinct porosity-permeability trends in a turbidite reservoir.

- An understanding of the lateral connectivity and heterogeneity of facies, which, in turn, enables an understanding of areal displacement efficiency (i.e. horizontal displacement within the reservoir layer in the x and y directions).
- An understanding of the structural framework and the fault network in terms of lateral barriers and/or baffles.
- The incorporation of pressure data and its tight integration with reservoir engineering and production geophysics.
- Understanding all aspects of rock and fluid properties essential to storage (in situ) characterization and fluid flow.

3.2 AQUIFER CHARACTERIZATION USING GEOPHYSICAL INFORMATION

As discussed in Chapter 2, geophysical techniques are used to extract relatively detailed subsurface information in 3D and at the local scale (field or prospect). In particular, seismic interpretation—which serves to characterize the geometry and quality of reservoirs and seals—is an integral function of the geoscientist. The combination of the reservoir and seal defines a container. We stated earlier that during seismic interpretation, much attention is focused on understanding the reservoir, its depositional environment and its structural history. Similar efforts can be focused (although this is not usually the case in the petroleum industry) on understanding the quality of top seals and the state of stress in a given area. Indeed, seismic velocities are quite responsive to stress vectors (Flemings 2021).

52 Integrated Aquifer Characterization and Modeling

The following five sections of this chapter present specific examples sourced from our experience in the petroleum industry that cover various aspects of aquifer characterization at the local scale.

3.2.1 Internal geometry of the aquifer: stratigraphy

When they are visible and can be found, the boundaries of an aquifer's areal extent are defined using seismic. Seismic also serves to assess the thickness of the aquifer and its potential internal complexity. Internal compartmentalization is often caused by faulting (see Section 3.2.2) and/or by stratigraphic complexity within the reservoir interval.

Figure 3.14 shows a seismic example from Australia's northwestern shelf, depicting the net reservoir HC thickness in a fluvial (channelized) environment. Fluvial environments are known for having discontinuous, very unpredictable geometries. The data on the left is a seismic amplitude extraction within a depth window corresponding to the entire reservoir interval. High amplitudes shown in red and orange reflect thick channels, while low amplitudes depicted in light greens and blues reflect very thin or absent HC reservoir. The analysis/interpretation is based on the fact that, given these sands' rock properties, seismic amplitudes should be approximately proportional to the following:

$$h \times \phi \times S_{HC} \tag{3.1}$$

where
- h is net vertical sand thickness,
- ϕ is porosity, and
- S_{HC} is HC saturation.

When either the net thickness, porosity or HC saturation increases, so do the seismic amplitudes. For more on this type of analysis, see Connolly et al. (2002) and Connolly (2007).

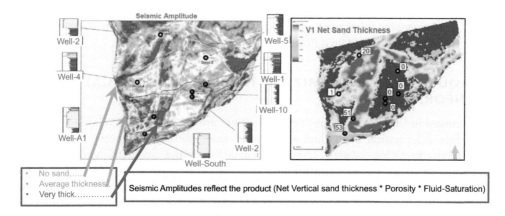

Figure 3.14 Seismic example reflecting the net reservoir HC thickness in a fluvial (channelized) environment (from Blangy 2020).

A similar analysis can be carried out—albeit with less sensitivity—within the water leg. In this case, seismic amplitudes correspond to the product of thickness times porosity at full water saturation.

It is important to remember that seismic velocities (as well as amplitudes) are affected by pressure, so all these "static" analyses from 3D assume a constant effective pressure. It is equally important to keep in mind that significant uncertainties likely remain regarding the full areal extent of aquifers owing to the high cost of 3D seismic and to the fact that, even when already acquired by O&G, seismic coverage most likely targets areas with HC potential—viewed as economically relevant—rather than aquifer definition—currently still considered of low economic relevance.

3.2.2 Internal geometry of the aquifer: faulting

One of the tasks of the geophysicist is to define the limits of the container by creating integrated container edge (ICE) maps. ICE maps focus on container definition. A conceptual example of a trap formed by two faults is shown in Figure 3.15. The interpreter first maps the faults, the top of the reservoir zone (in blue) and the bottom of the reservoir zone (in green). Three trap edges are identified: a top seal, a western delimiting fault (in purple) and an eastern delimiting fault (in blue). A fourth edge (for HCs only) would be formed by the oil–water contact (OWC or WOC). In order to calculate the volume enclosed above the OWC—called the gross rock volume (GRV)—this fourth edge can be used to trim the overall volume down to the limit of the base reservoir. GRV computations do not take into account the properties of the rock contained in the trap.

A good example of using seismic in order to assess volumetrics is the study published by Swinburn et al. (2011) supporting reserve estimates in the Gorgon field offshore Australia. The study shows "type-III" sands—characterized by deep red, negative seismic amplitudes or low acoustic impedance when gas-saturated—exhibiting much dimmer amplitudes (light

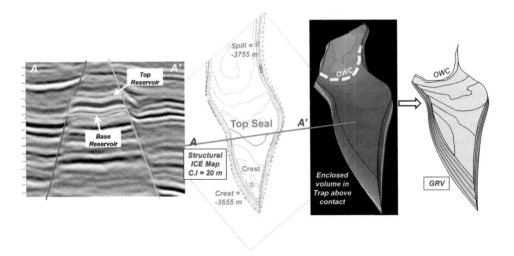

Figure 3.15 Conceptual example of a GRV construction through ICE identification (from Blangy 2020).

54 Integrated Aquifer Characterization and Modeling

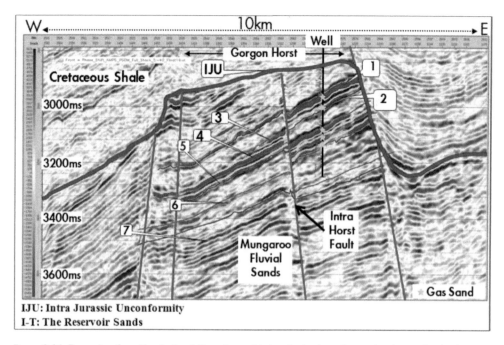

Figure 3.16 Example of aquifer limits delineation, with key faults, based on seismic amplitudes in a gas field (adapted from Swinburn et al. 2011).

red) within the aquifer. Figure 3.16 shows their seismic interpretation along a W–E direction. The seismic data type is a far-stack, 90 degree phase-shifted amplitude, which is similar to a pseudo-impedance. The corresponding amplitude relationship extracted around sand interval 3 is shown in map view on the right-hand side of the figure. The integrated container edges are clearly delineated by the color extractions on the right-hand side of Figure 3.16, while the aquifer (in blue) shows some of its internal geometries.

Furthermore, note the faults (shown in red on the seismic) that may compartmentalize the reservoir. Assessing connectivity across faults can be complex, as this requires more than fault identification, i.e. the task entails the need to characterize whether the fault acts as a seal, baffle or is non-compartmentalizing. This can be approximated through fault juxtaposition as described in Section 3.1.3. In Figure 3.16, the main fault in the middle of the Gorgon horst (labeled "Intra Horst Fault") likely acts as a seal at this location because it appears to be showing sand-against-shale juxtaposition. However, a 3D analysis is required in order to determine whether the fault tips out along strike (N–S direction) within the structure. If the fault tips out, then sand is juxtaposed to sand at this location and the fault is likely a conduit for fluids. As we discussed in the structural geology section, other factors—shale gouge, grain cataclasis, local stress with respect to the orientation and dip of the fault—affect the fault's ability to compartmentalize the segment. Clearly, any dynamic data that captures the impact of production with time (such as pressure or fluid saturation changes coming from wells and/or 4D surveillance data) is invaluable in validating predictions of fault transmissivity. We will discuss this in more detail in Section 3.2.6.

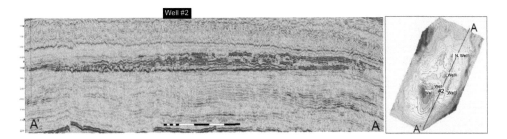

Figure 3.17 An example of a seismic GWC-calibrated "flat spot" displayed along the strike of a gas field. The contact appears fairly continuous and relatively flat (+/- 2 m depth difference) over a distance of ~20 km (from Blangy 2020).

3.2.3 Gas–water contact identification

Fluid contacts often carry a significant amount of uncertainty until they are intersected by wells and evaluated through well logging and/or fluid sampling. However, several indirect techniques—such as remote sensing—can, under the right circumstances, be used to approximate the location of contacts.

Figure 3.17 shows an example of a seismic "flat spot" calibrated to a gas–water contact (GWC) and displayed along the strike of a gas field. All five wells shown in the map on the right-hand side were tied to the seismic (via seismic synthetic seismograms) and a robust rock physics model (RPM) indicates that gas-saturated sands are low impedance (in deep red) while wet sands are much dimmer. Furthermore, the gas–water interface is predicted as a strong positive acoustic contrast (blue reflection appearing as a flat spot), as shown. The contact appears to be fairly continuous over a large distance (almost 20 km) and relatively flat, with a depth difference of +/- 2 m over the area. Furthermore, the flat spot was further shown to follow structural contours (not shown here), thus indicating a high-confidence velocity model (based on full-waveform inversion, FWI) and accompanying depth conversion. Once calibrated to the wells, this seismic was used as a reliable technology to map the GWC away from the wells. We can conclude that a common aquifer is likely present for the reservoirs in this field.

3.2.4 Oil–water contact identification

OWCs sometimes manifest as seismic "flat spots," but only seldom because the impedance contrast between an oil-saturated rock and a water-saturated rock is not as large as the impedance contrast between a gas-saturated rock and a water-saturated rock (see Section 3.2.3). However, other, more subtle seismic attributes (such as amplitude and/or phase changes) are often used by O&G to help define OWCs. Most often, these types of seismic attribute changes need to be interpreted on the basis of a RPM.

Figure 3.18 is a "chair display" (the intersection of two seismic lines at a right angle) from an anonymous oil field showing a seismic amplitude extraction draped along the reservoir's depth representation. Note the relatively sharp amplitude change as indicated by the white arrows. The amplitude contrast appears to show strong structural conformance (just below

56 Integrated Aquifer Characterization and Modeling

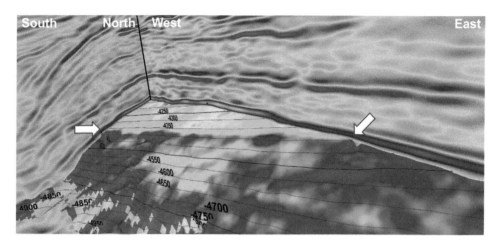

Figure 3.18 3D perspective of a seismic amplitude change extracted from a reservoir interval and confirmed by wells as an oil field OWC (from Blangy 2020).

the 4350 contour) and has been calibrated to the OWC of a well drilled elsewhere in the field. Also note the subtle amplitude changes (yellows to greens) within the oil leg, indicating possible variations in reservoir facies.

Oftentimes, the identification of fluid contacts is enhanced through the use of 4D seismic, which is a repeat survey pre- and post-production. Thus, 4D (or time-lapse) seismic measures the response of the subsurface (encompassing reservoir and seal pairs as well as overburden) to production processes that include changes in pressure and fluid movements. For example, Figure 3.19 shows a 3D perspective of a reservoir interval in the GoM Atlantis field with a complex salt overburden to the north. The draped display is a 4D amplitude difference between 2005 and 2019. Note the movement of the original water contact (the dark blue ribbon), indicating a hardening effect. The deepest edge of the ribbon is the original water contact, while the highest edge of the same ribbon is the current OWC. At this field-wide scale, the reservoir appears to be responding "normally" to production, with peripheral water movement toward the center of the anticline as oil is produced. When zooming in to the next scale down, i.e. at the segment scale (see Chapter 2), local complexities in fluid movement—with barriers and baffles—are likely to appear. More examples of production effects at the segment scale are discussed in Section 3.2.6.

3.2.5 Rock-physics-based seismic inversion for facies classification

Seismic inversion allows the transformation from ***reflectivity***—the native domain where seismic is recorded, consisting of seismic reflection coefficients arising from layer boundaries or interfaces—to ***impedance***, i.e. the elastic layer properties of the interval between seismic reflections. Because seismic inversion unlocks the characterization of quantitative rock and fluid properties, it is a key tool in seismic reservoir characterization. Seismic inversion usually includes reservoir measurements such as those from well logs and cores

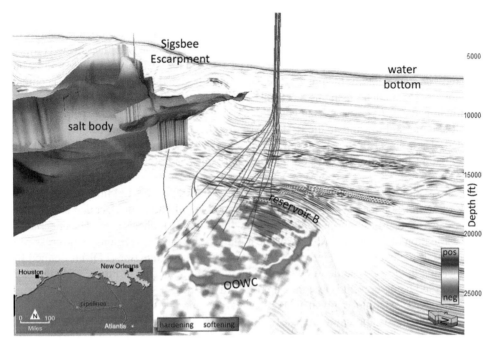

Figure 3.19 3D perspective of the Atlantis Field showing the main reservoir surface, seismic section and complex salt in the overburden. The draped surface shows a 4D difference between 2005 and 2019 extracted from one of the key reservoir intervals (reservoir B). The blue lines indicate the location of the wells currently producing from the platform (from Van Gestel 2021).

used to calibrate its results. To compensate for the band-limited nature of recorded seismic, seismic inversion often incorporates very low frequencies extracted (a) from pre-stack data migration- (PSDM) or FWI-derived seismic velocities, and/or (b) from regional compaction trends estimated via well logs. Seismic inversion can be performed pre- or post-stack through various methods. Elastic (pre-stack) inversion tends to be computationally intensive and will not be addressed here beyond noting that workflows include deterministic (i.e. sparse-spike), stochastic/statistical and wave equation-based methods (Contreras et al. 2020). The calibration of seismic inversion results to well control is key. For a good example of a simultaneous pre-stack inversion workflow and its calibration to wells, see Singleton and Keirstead (2011).

A typical quantitative seismic interpretation workflow includes a preliminary seismic structural-stratigraphic interpretation that yields a geometric framework then used as a reference for the seismic inversion. The output from the seismic inversion enables the extraction of elastic properties, which in turn are used to classify seismic facies (i.e. reservoir facies or fluid type etc.) through a Rock Physics Template (RPT) and/or a petro-elastic model (PEM).

A RPT or a PEM is a key interpretation tool serving as a bridge between geology and geophysics and enabling these disciplines' quantitative integration, thus unlocking seismic reservoir characterization (SRC). As shown in a number of published case histories, the key to SRC is to map the main geological/petrophysical facies of interest into the seismic elastic space. Once the elastic space is defined in a way that allows for the classification to be established, what remains to be done is the identification of points in confidence regions on the cross-plot and their projection back onto the 3D seismic volume with the selected properties. There are many elastic parameters that can be selected for the analysis of cross-plots, but depending on the quality of the pre-stack seismic and the signal-to-noise ratio (S/N) content, we generally recommend using as a starting point either the Poisson-ratio or Vp/Vs versus acoustic impedance (AI) space (Ball et al. 2018). Regardless of which elastic parameters are used, it is important to validate the RPT/PEM through a comparison between measured and modeled elastic logs. Further local refinements to the RPT can be introduced through the differentiation between well and poorly cemented zones as described by Avseth and Skjei (2011).

The cross-plot on the left-hand side of Figure 3.20 shows the effect of lithology and fluids (water saturation) on the elastic properties of the rock, highlighting the fact that shales have similar elastic properties to brine sands, though with a slightly higher Vp/Vs. The uncertainty in the estimation of facies (or lithologies) due to the overlap between regions in the cross-plot is captured through a Bayesian inference; a probability density function (PDF) is defined for each location in the 3D seismic. As displayed on the right-hand side of Figure 3.20, the likelihood of sand being present is assigned a probability value.

Another RPT example is given in Figure 3.21 (Pendrel and Schouten 2020), where a Bayesian inference is applied to the elastic cross-plot of Vp/Vs versus AI in order to derive the occurrence probability for each facies at every location in the 3D volume. The design breakdown of this study's facies template is shown on the left-hand side of Figure 3.21, where the three key facies chosen—shale, silt, pay sand (which contains HC fluid)—are identified. The elastic PDFs are shown as three ellipses constructed by means of a RPT shown in light blue lines. The output (right-hand side of the figure) illustrates the most likely facies at each location within the seismic grid. An uncertainty and/or confidence index is associated with each prediction in space. As Pendrel and Schouten emphasized, facies-based Bayesian inversions incorporate prior information from wells and account for uncertainty in 3D, constituting the current state-of-the art approach to seismic inversion.

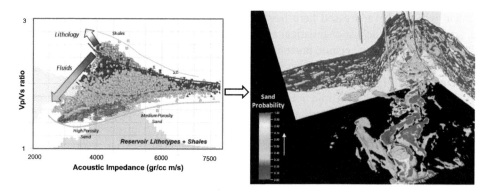

Figure 3.20 The definition of petro-elastic regions based on a RPT (left-hand side) generates the sand probability map shown on the right-hand side (modified after Webb et al. 2020).

Aquifer description and characterization 59

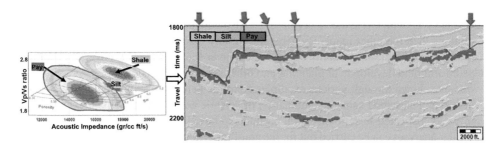

Figure 3.21 A RPT is used to guide the definition of elastic probability distribution functions for three key selected facies. i.e. shale, silt and pay sands. PDFs help predict which facies is most likely present at each location within the 3D seismic grid (modified from Pendrel and Schouten 2020).

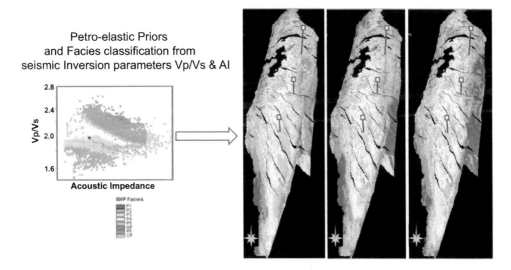

Figure 3.22 Facies inversion example using a stochastic inversion for facies. Seismic inversion results were calibrated to three wells. Data from wells and cores was utilized to define characteristic properties by facies. The three maps on the right represent equally probable distributions of reservoir facies for a particular layer within the 3D volume (modified from Blangy 2020).

Figure 3.22 depicts a third example of facies inversion, one using a stochastic approach. Seismic inversion data was used to define key facies in a petroelastic space (Vp/Vs versus AI). Data from three wells (displayed on the right-hand side)—including core descriptions—was used for calibration, supporting the assignment of elastic characteristics to each individual facies in the cross-plot. Multiple facies were assigned properties from the core/log observations: facies 2 and 3 were found to correspond to coarse- and fine-grained amalgamated turbidite sands, facies 4 and 5 are sand- and mud-prone heterolithics, facies 6 is mudstone, facies 7 consists of remobilized deposits and facies 8 consists of cemented lithologies or carbonate marls. The three maps on the right-hand side display equally probable facies distributions for a particular geological layer. Note the dominance of facies 3, 4 and 5 for this particular layer, which is at the reservoir level. Also note that the left-most map appears more continuous than the other maps. By generating many possible facies distributions, one can define an overall facies uncertainty by geological layer for this field. This overall uncertainty helps inform field development planning and other economic decisions.

We have shown through various examples from the petroleum industry that facies can be identified and classified using a combination of seismic and well data. This level of analysis requires sophisticated seismic inversion products to be calibrated to well data.

Similar analyses can be carried out for the aquifer portion of geological deposits. Generally speaking, differentiating sub-facies types within aquifers by using seismic reservoir characterization is more difficult than doing the same within a HC leg because aquifers' seismic displays a higher velocity due to fluids and pressures.

Nevertheless, like HC reservoirs, connected aquifers will exhibit seismic variations (in velocity, amplitude and frequency) due to facies, porosity and/or pressure changes. Some of these variations within aquifers were seen in Figure 3.18 (the transition from blues to purples) and in Figures 2.10 and 2.11 (where the aquifer shows complex channel geometries) in Chapter 2. These variations within the aquifer are likely indicative of changes in facies as well as changes in storage properties.

3.2.6 Production geophysics and surveillance

4D seismic, when used in conjunction with pressure histories from well data, is a key tool for the "monitoring and verification" of production processes. An integrated surveillance approach incorporating all subsurface disciplines can both generate an account of production-induced subsurface volumetric changes and verify it at the segment scale.

When performing seismic surveillance, a quantitative approach to the interpretation of seismic attributes (usually based on pre-stack inversion and solid rock physics modelling, RPMAC, as discussed in section 3.2.5) is recommended, so as to maximize the detection and understanding of subtle changes in the 4D. Pre-stack inversion involves both pre- and post-production seismic data. In order to make results coming from multiple datasets more robust, interpreters have used a 4D "simultaneous elastic inversion," inverting these datasets together rather than sequentially. Many authors have shown that 4D allows for (a) more detailed seismic interpretations of internal connectivity, and (b) a superior understanding of production processes achieved through the observation of reservoir changes due to fluid movement. These changes can be observed in fluid pressures (Johnston and Laugier 2012) and saturations (Berthet et al. 2015, Nasser et al. 2016, Blangy et al. 2017, Aikulola et al. 2020), and sometimes also in reservoir compaction (De Gennaro et al. 2017).

Figure 3.23 (modified after Blangy et al. 2017) shows the 4D seismic response in a turbidite reservoir from a producing field offshore West Africa. The attribute chosen for visualization is the difference in acoustic impedance between 2010 and 2014. Hardening (shown in light blue) is due to water encroachment while softening (in red) is due to over-pressuring, as confirmed by nearby producing wells. The original OWC—related to the main aquifer in the area (at 1,980 m)—is shown as a flat surface for reference. Water is being injected in the field just above the OWC for pressure maintenance. An injector directs water in a N–S direction, as shown by the arrows. Note the rather tortuous hydraulic connection between the

Aquifer description and characterization   61

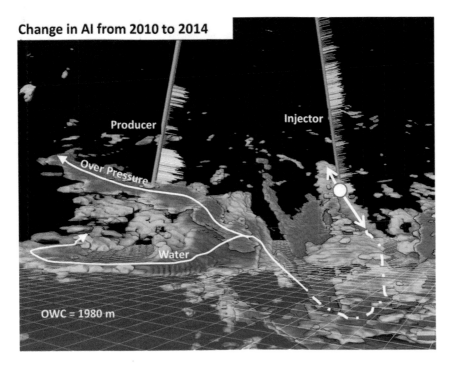

Figure 3.23 Visualization of an oil reservoir and its aquifer from a field offshore West Africa, with the path taken by water injected via a dedicated well (from Blangy et al. 2017).

injector–producer pair. It is interesting that the water makes its way to the nearby producer in a sinuous fashion, following the distribution of preferred permeability in the channel system. The water reaches the aquifer but it is also diverted in part through another channel, which becomes overpressured over time. It would be very difficult—if not impossible—to predict the path of the fluids from injector to producer without the 3D visualization from the 4D seismic.

The quantitative interpretation of 4D seismic enables a significant new understanding of fluid movement through time while also helping decipher the evolving connectivity of bodies of water in producing fields. This understanding supports the evaluation of fault behavior—i.e. whether faults leak/seal over time—and serves to highlight undrained areas.

Figure 3.24 illustrates the change in acoustic impedance occurring between 2005 and 2015 within a reservoir interval in the southern portion of a producing field. Hardening (shown in blue) is due to water encroachment. The white arrow shows the original OWC (2005), while the yellow arrow indicates the new OWC (2015). Note the excellent overall conformance of water movement to the structure in that part of the field. It appears that the main aquifer responds uniformly to production during the observed duration. In addition, some of the faults shown in black clearly compartmentalize the reservoir (arrows in light and dark green) while other faults appear to behave more like baffles, with minimal change observed across them. Also note, in the eastern part of the field (red arrow), a significantly more pronounced area of water advancing in an up-dip direction.

62 Integrated Aquifer Characterization and Modeling

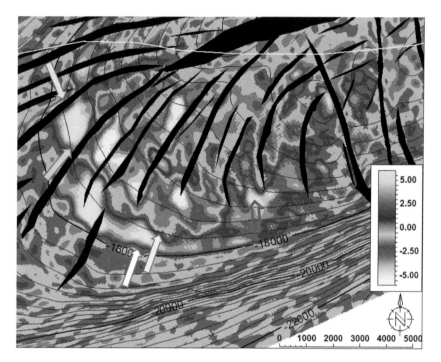

Figure 3.24 Difference in acoustic impedance between 2005 and 2015, within the reservoir interval of a producing field. (Figure, courtesy of Kelly Wrobel).

Together with reservoir engineering models and simulations, 4D seismic can serve as input for *aquifer strength* characterization because it facilitates the estimation of volume change over time. This aquifer strength assessment method deploys a mass-balance approach to fluid volumes and takes into account the evolution of fluid pressures through time. This topic is further developed in subsequent chapters.

Let us now turn to economic considerations for seismic-based monitoring: While 4D seismic is a useful tool for assessing production processes, it can be relatively expensive to run 4D surface seismic at frequent intervals. Lower-cost nonintrusive technology has been developed and tested that involves the use of (a) fiber optic cables permanently installed in wells, and (b) distributed acoustic sensing (DAS) for monitoring production processes. Kiyashchenko et al. (2020) insightfully compare 4D DAS vertical seismic profiles (VSP) to 4D ocean bottom node (OBN) seismic in a GoM field. The field is submerged for the purpose of helping recovery, and 4D seismic is used to track the movement of the water front (Figure 3.25). Contours (not labeled) indicate a homocline with up-dip in a western direction. Figure 3.25a shows the water injection front tracking via 4D OBN seismic amplitude differentiation between 2007 and 2017. The trace of the injector well is shown in red. Figure 3.25b shows path predictions—developed using reservoir simulation—for water injected from 2015 to 2017. A repeat OBN survey acquired in 2018 shows the hardening of the reservoir due to water injection over the three-year period from 2015 to 2018 (Figure 3.25c). During

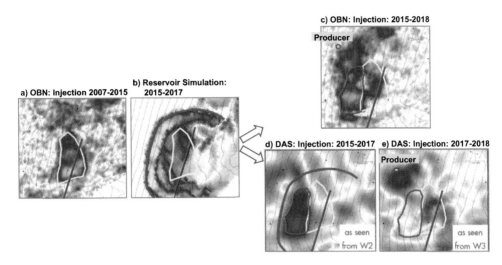

Figure 3.25 (a) Water injection front tracked via 4D OBN (2007–2017); (b) reservoir simulation path predictions for injected water (2015–2017); (c) repeat OBN showing reservoir hardening due to water injection (2015–2018); (d) repeat DAS-VSP showing a first episode of water movement in an up-dip direction (2015–2017); (e) repeat DAS-VSP showing a second episode of water movement in an unexpected direction (2017–2018) (modified after Kiyashchenko et al. 2020).

that time, the water moved up-dip from the injector while also encroaching on the producer (green dot). Lastly, the (d) and (e) segments of Figure 3.25 show repeat DAS-VSP surveys from 2015–2017 and 2017–2018. The DAS surveys were acquired on a more frequent basis and show two distinct episodes of water movement: one up-dip (2015–2017), followed by water flow through a preferred propagation path along-strike to the north and toward the producer (2017–2018).

Kiyashchenko et al. (2020) showed convincingly that—despite its limited illumination (less extensive lateral coverage) and despite exhibiting more coherent noise when compared to OBN data—the 3D DAS-VSPs, acquired simultaneously in multiple wells, allowed for the timely detection of the unexpected pattern in the water-sweep.

While 4D seismic is a proven technology for the "monitoring and verification" of production processes, DAS 4D monitoring is a lower-cost emerging technology that is likely to gain ground in the near future for reservoir surveillance and management. Both seismic monitoring techniques (be it surface seismic-based or well-based via DAS) lend themselves to the characterization of injection processes and to the monitoring of CO_2, methane, hydrogen and/or waste liquid sequestration sites (see Chapter 9).

3.3 AQUIFER CHARACTERIZATION USING PETROPHYSICAL INFORMATION

Petrophysical analysis and interpretation is a key discipline that leverages the information collected in wells to characterize encountered rock and fluids.

The following topics will be discussed next:

- standard petrophysical interpretation
- transition zone
- paleo zone and trapped HC
- tar mat

3.3.1 Petrophysical interpretation

Table 3.2 shows traditional open-hole wireline well logging tools and their application. Most O&G applications acquire a combination of these logs and interpret them based on specific reservoir characterization needs. The logs are classified according to lithology, porosity, saturation and imaging. Other useful logs not included in the summary table below include caliper logs (used to estimate borehole diameter and elongation), temperature logs, directional logs and formation testers (for pressure estimation and fluid sampling).

In addition to open-hole logs, cased-hole and cement bond logs (not shown in Table 3.2) can provide useful information. For example, pulsed neutron capture (PNC) tools are a primary

Table 3.2 Major open-hole wireline logs, their classification and their main application

Log classification	Log type	Application
Lithology logs	Spontaneous Potential (SP)	Lithology indicator for correlation with other logs and other wells Provide Rw (water resistivity) Determine boundaries between permeable and non-permeable formations and provide an indicator of shaliness
	Gamma Ray (GR)	Provide lithology data for correlation with other logs and other wells Determine boundaries between permeable and non-permeable formations and provide an indicator of shaliness Enable clay typing (Spectral Gamma)
Porosity logs	Sonic (including Sonic scanner and Dipole Shear)	Provide porosity data Provide indirect data to distinguish between oil and gas Establish and monitor state of stress Provide mechanical properties
	Neutron	Provide accurate lithology and porosity determination Provide data to distinguish between oil and gas Provide porosity data for water saturation determination
	Density	Provide accurate lithology and porosity determination Provide data to distinguish between oil and gas Evaluation of formation density Provide porosity data for water saturation determination

Table 3.2 Cont.

Log classification	Log type	Application
Saturation logs	Various types of Resistivity logs (mostly Induction logs, but also Normal, and Laterologs) and Microresistivity logs	Determine the thickness of the formation with accuracy Provide an accurate value of formation resistivity Provide an indication of formation pressure, hydrocarbon saturation and producibility Provide information for correlation purposes Provide dip information (in case of the Dipmeter)
	Nuclear Magnetic Resonance (NMR)	Analysis of porosity, permeability and fluid volumes Details of fluid saturation (bound fluids, free fluids) Analysis of moveable fluids
	Dielectric scanner	Water salinity Archie's exponents M and N Cation exchange capacity (CEC) and detailed clay volumes Detailed Mineralogy
	Rt scanner	Directional resistivity for formations with low resistivity contrast Accurate thin-bed analysis Acurate fluid saturations in laminated, anisotropic and/or faulted reservoirs
Spectroscopy tools	Litho Scanner, Element Capture Spectroscopy	Matrix composition, through elemental mineral fractions Estimation of Lithology and Porosity Determination of total organic carbon (TOC) in shales
Imaging logs	Sonic Imaging	Analysis of stress, fractures Establish and monitor state of stress Imaging of the borehole wall: Borehole stability
	Resistivity Imaging	Analysis of stress, fractures Establish and monitor state of stress Dipmeter applications: dip and azimuth of bedding and faults Imaging of the borehole wall: Borehole stability

means of evaluation for reservoir monitoring and surveillance during field production. In particular, pulsed neutron spectroscopy and full waveform acoustics are promising methods for measuring lithology, porosity and saturations through casing, even in the presence of freshwater. In general, cased hole logs (capture cross-section, sigma) are used to estimate the ability of fluids to flow through the formation.

When core data is available, it is important to calibrate log-derived rock and fluid properties to those observed on cores. The use of core information is essential to ground truthing issues pertaining to reservoir complexity observed at the borehole scale. A typical example of this is shown in Figure 3.26, which compares core versus log porosity and permeability.

Another important application of log analysis is the establishment of cut-offs for various zones. Chosen cut-offs define criteria for reservoir quality and net thicknesses. As can be inferred from Figure 3.26, the various cut-offs set on input control the output volumetrics/net quantities for lithological units of interest and their continuity.

66 Integrated Aquifer Characterization and Modeling

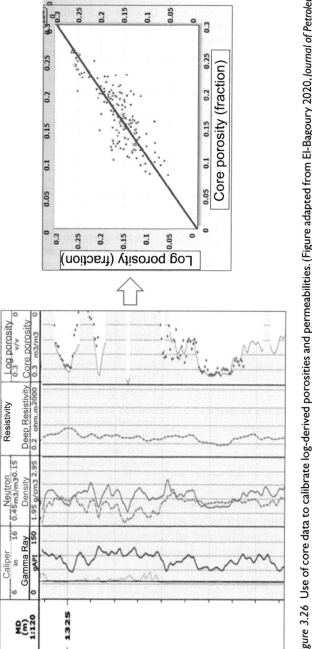

Figure 3.26 Use of core data to calibrate log-derived porosities and permeabilities. (Figure adapted from El-Bagoury 2020, Journal of Petroleum Exploration and Production Technology. Reprinted by permission of the AAPG whose permission is required for further use).

Petrophysical Cutoffs - Discussion

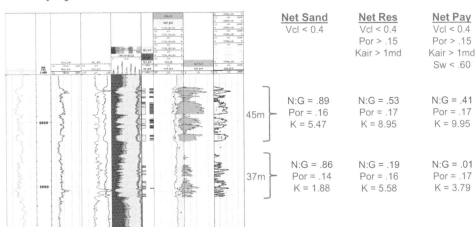

Figure 3.27 Use of core data to calibrate log-derived porosities and permeabilities within a HC zone.

In Figure 3.27, the gamma ray response is used to define "net sand" as sands with a clay volume (Vcl in the cited figure) of less than 40%. Net sands are then defined as "net reservoirs" (Net Res in the cited figure) through minimum cut-offs for porosity and permeability. Lastly, since we are dealing with a hydrocarbon reservoir in this example, a HC saturation cut-off is set at greater than 40% (water saturation or $S_w < 60\%$) to establish Net Hydrocarbon Pay. The average properties within the net intervals defined in Figure 3.27 as "net sand," "net reservoir" and "net pay" are different, increasing progressively as a result of the cut-offs. Note that Net-to-Gross (NtG or N:G) is a property that is also a function of the cut-offs (and scale).

The O&G practice of applying different cut-offs during log analysis of pay zones is directly applicable to the characterization of aquifer zones.

As described in Section 3.1.4, porosity-permeability trends are often controlled by geological facies. Porosity is a key storage property, while permeability is a key indicator of flowability. In general, both porosity and permeability "trend" together. However, a reservoir with good porosity does not always flow at desired rates. One way of categorizing heterogeneity in a reservoir is to use the Lorentz plot, which involves plotting fractional total flow capacity (permeability times thickness, i.e. k*h) as a function of fractional total storage (i.e. porosity times thickness, i.e. ϕ*h or "Por * h" in Figure 3.28a below). A homogeneous unit will plot close to the diagonal, while a unit with significant heterogeneity will plot above the diagonal (Figure 3.28a). The Lorentz plot is also used to arrive at the correct grid size for upscaling the numerical model. A comparison of the actual log data and the grid data for a coarse-scale versus fine-scale model is shown in Figure 3.28b.

In addition to characterizing the properties of lithologic intervals (flow units, aquifers) by incorporating different tools and subsurface disciplines, it is important to describe and quantify the nature of the interface between the aquifer portion and its oil or gas counterpart within a HC reservoir. Petrophysical tools provide a good estimate of the transition of fluids, which are usually segregated by gravity, at least at the local scale within a given segment. Once this

68 Integrated Aquifer Characterization and Modeling

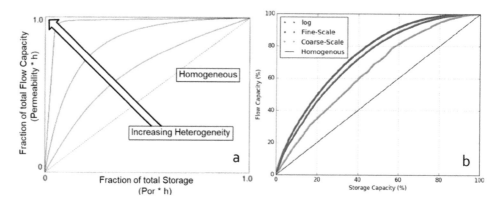

Figure 3.28 Use of Lorentz plots to define and verify reservoir heterogeneity (a) and its modeling (b).

interface is described, its impact on the overall connectivity of the aquifer can be assessed. The next section addresses ways of estimating aquifer connectivity at the interface.

3.3.2 Transition zones

A *transition zone* is defined as a region where S_w changes progressively from 100% ("native state" of the formation) to a region of increasing HC saturation and up until irreducible water saturation (S_{wirr}) is reached. If a petroleum reservoir is perforated at and produced from a transition zone, it will co-produce a mixture of HCs and water.

A review of *capillary pressure* concepts is timely at this stage of our discussion since they underpin the presence or absence of a transition zone.

3.3.2.1 Capillary pressure

Capillary pressure (P_c) is defined as the difference between the pressure in the non-wetting phase and the pressure in the wetting phase across the fluid–fluid interface (i.e. the interface between the *free water level*, FWL, and the OWC). Depending on the heterogeneity of the grains, during primary drainage—for instance when oil or gas is displacing the saturated rock—a certain initial pressure is required for the injectant to enter the saturated rock. This is called *threshold or entry pressure* (see Figure 3.29). Then, as the drainage process continues, funicular regions are created where injectant saturation increases within pores. Finally, in the pendular region, S_w reaches close to S_{wirr} and an oil column free of water production develops.

Notice that the efficacy of an oil or gas field's seal rock and the potential HC column height it can sustain can be characterized by examining the cap rock capillary entry pressure.

3.3.2.2 FWL versus OWC

A transition zone (known as the FWL) develops at the interface between an aquifer with S_w at 100% and oil or gas reservoirs at their initial HC saturation (see Figure 3.30). The actual WOC or GWC occurs at a certain height above the FWL, where the capillary pressure is equal to entry pressure. The FWL is detected from modular dynamic tester (MDT) data,

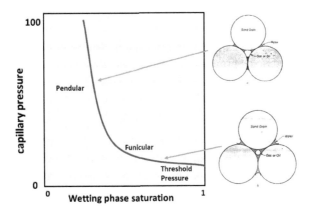

Figure 3.29 Capillary pressure curve (modified after Leverett 1941).

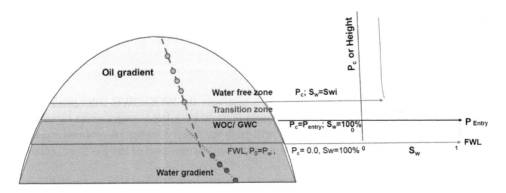

Figure 3.30 Aquifer–reservoir interface.

while the WOC is derived from core and resistivity log measurements. In most cases, FWL and WOC are the same because the entry pressure is zero (due to large grain size). However, in lower quality rocks there could be a significant depth difference between the two, so the assumption that FWL and WOC are equal may lead to the overestimation of the volume of oil or gas originally/initially in place (OOIP or GIIP/OGIP). While FWL is always flat, WOC or GWC may be tilted in some cases because entry pressure is a rock property. If FWL measurements differ within the same reservoir, this is due to compartmentalization.

While S_w in a transition zone increases with depth, this does not correspond to an increase in water production because relative permeability controls how much water can flow in this zone.

Larsen et al. (2000) classified the reservoir transition zone into three categories:

- homogeneous (reservoir sand without thin layers of shale)
- layer dependent (thin shale layer occurring in reservoir sand)
- heterogeneous transition zones (multiple thin shale layers occurring within reservoir sand)

The first step in any transition zone analysis is procuring the rock capillary pressure data. Then, other data such as HC–water contacts, rock matrix type, pore throat structure and transition zone height can be determined.

Abiola and Obasuyi (2020) carried out a transition zone analysis on the Stephs field in Nigeria using a methodology proposed by Goda and Behrenbruch (2011) for estimating capillary pressure from well logs and 3D seismic data. No core data was available for this analysis. Travel time against depth was generated using 'Stephs' field checkshot data. A synthetic seismogram was generated using Butterwort wavelet at zero phase sampled at an interval of 4 milliseconds (ms) to establish a link between the seismic and the well. Figure 3.31 displays the final estimated transition zone and the capillary pressure model for reservoir B. It is unclear how the seismic data was used to generate the capillary pressure map. Facies data and estimated capillary pressure data could potentially have been used to generate the map illustrated in Figure 3.31 below without using seismic data. The transition zone is evident on the capillary pressure log and on the model as the color progresses from deep blue to green and finally to light blue, with capillary pressure values ranging from 6 to 18 psi (labeled D). The zone marked in red with values ranging from 18 to 35 psi displays a high capillary pressure, which is indicative of zones fully saturated with HC (oil); this is evident on the eastern part of the model and in the region labelled "E" on the capillary pressure log and model (Abiola and Obasuyi 2020).

3.3.3 Paleo zone

Once oil and gas columns are established in a reservoir, the WOC/OWC will remain static. However, in many reservoirs, due to depletion (caused by a leaky reservoir or by production from surrounding wells), the aquifer will push the water into the HC column, creating a so-called "paleo" zone. In a gas reservoir, water will bypass part of the gas zone, creating trapped gas saturation (S_g).

In the case of high compressibility reservoirs, e.g. dry gas, the impact of an aquifer is usually detrimental because of the gas-trapping phenomenon. Gas relative permeability during the drainage process—i.e. when the reservoir is filled up with gas—is much higher than when gas is being displaced by water from an aquifer, which is an imbibition process. During drainage, the ***end-point residual S_g***—also called the ***critical S_g***—is typically 2–5% while the imbibition end-point saturation to gas (or "trapped" gas) is significantly larger—usually in the range of 20–40%. In some cases, such as mouldic limestones in the Qatar North field, trapped gas has been measured to be as high as 70%.

Aquifer influx in gas reservoirs, therefore, leads to gas trapping and a reduction of ultimate recovery. Since the compressibility of water is much lower than that of gas, an aquifer has to be much larger than an oil reservoir to have significant impact. If an aquifer is larger than twenty times the gas reservoir volume, water influx appears to act more like an "infinite" aquifer, whose impact also depends on reservoir geometry, aquifer proximity and properties, etc.—as described above.

With sufficient time, the volume of aquifer encroachment becomes a function of aquifer pressure depletion and of aquifer pore volume and fluid compressibilities, as shown in Figure 3.32. With good sweep, 0.3 HC pore volume (HCPV) of aquifer encroachment could provide 30% HC recovery factor (RF).

Aquifer description and characterization 71

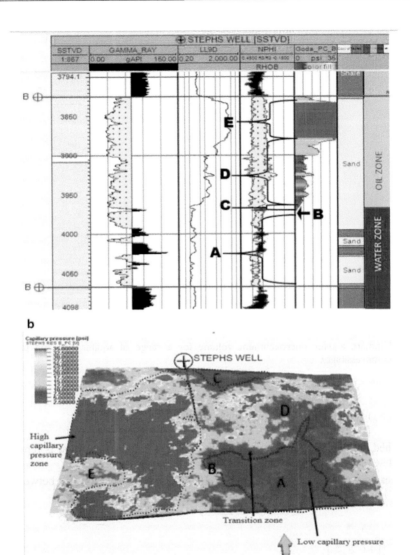

Figure 3.31 (a) Well section with capillary pressure log for reservoir B, where A is FWL, B is OWC, C is entry pressure, D is transition zone and E is the zone fully saturated with HC; (b) capillary pressure model for reservoir B, where A, B, D and E are defined the same as for (a) and C is low P_c zone (Abiola and Obasuyi 2020).

The published data for maximum trapped gas saturation versus porosity are shown in Figure 3.33.

Generally, trapped gas saturation (S_{gt}) depends on the initial gas saturation (see Kralik et al. 2000). In terms of CO_2 sequestration, we are mostly interested in trapping capacity, C_{trap},

$$C_{trap} = \emptyset S_{nwr} \tag{3.1}$$

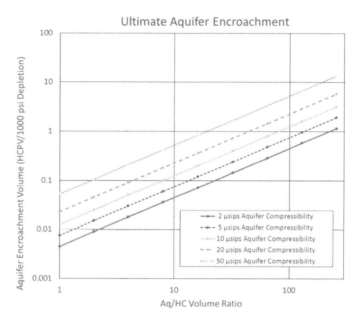

Figure 3.32 Ultimate aquifer encroachment volume for a range of aquifer size and pore volume compressibility.

where

S_{nwr} (e.g. S_{gt}) is trapped non-wetting phase saturation.

The published data on trapped gas saturation versus initial gas saturation is reproduced in Figure 3.34.

In pioneering work, Land (1968a, 1968b, 1971) derived a relationship between S_{gt} as a function of initial gas saturation in the rock:

$$S_{gt}^* = \frac{S_{gi}^*}{\left(1 + C S_{gi}^*\right)} \tag{3.2}$$

where

C is a constant and S* is normalized saturation.

$$S^* = \frac{S}{1 - S_{wc}} \tag{3.3}$$

The same relationship holds for trapped oil saturation as a function of initial oil saturation.

The presence of trapped oil (or gas) saturation impacts water relative permeability and, consequently, aquifer effectiveness. The relative permeability of water drops below 1 in the presence of trapped oil or gas, as shown in Figure 3.35. An example of reduction in water

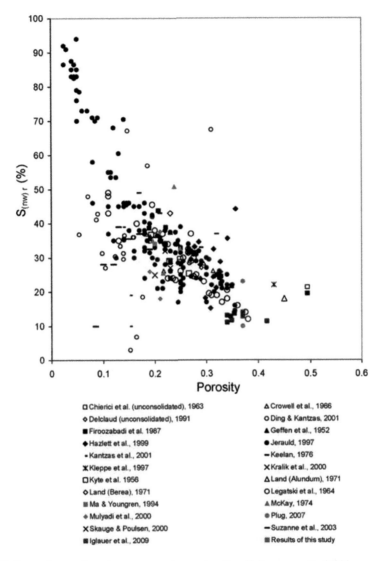

Figure 3.33 Dependence of maximum trapped gas saturation (S_{gt}) on porosity (Al Mansoori et al. 2010).

relative permeability versus porosity (due to trapped saturation, S_{gt}) is shown in Figure 3.36 (colored symbols represent different fields). Detection of trapped HC below WOC requires integrated core, log and test data.

Kheidri et al. (2016) demonstrated the presence of oil below the FWL in a United Arab Emirates (UAE) offshore oil field based on the following (see Figure 3.37):

- The log data showed the presence of oil below the transition zone, with high oil saturation below the FWL.

74 Integrated Aquifer Characterization and Modeling

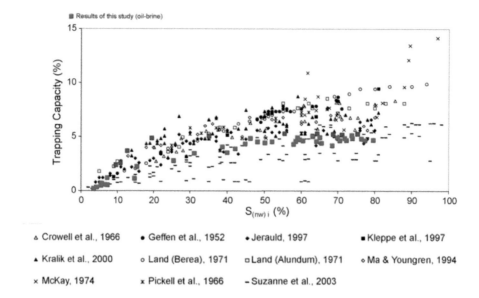

Figure 3.34 Trapping capacity versus initial non-wetting phase saturation (Al Mansoori et al. 2010).

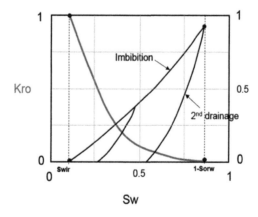

Figure 3.35 Water relative permeability.

- The plots of S_w versus depth for all the wells drilled in the studied field showed the presence of oil below the FWL.
- Drill stem tester (DST) results at the base of the main oil column showed the production of formation water (260,000 ppm NaCl) along with oil, indicating high water mobility—a characteristic feature of transition zones. High oil saturation can be identified below the tested interval, indicating the presence of "bypassed" oil.
- The description of the core below the FWL showed the presence of heavy oil staining.
- The plot of core plug report data versus depth showed the presence of oil below the FWL (see Figure 3.38).

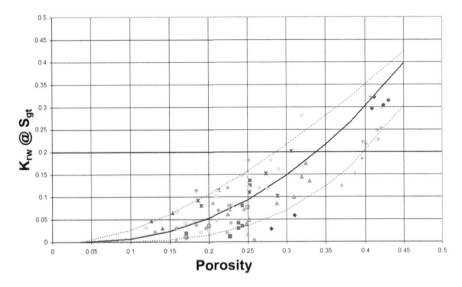

Figure 3.36 End-point water relative permeability versus porosity.

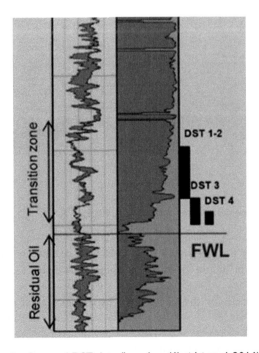

Figure 3.37 Water saturation logs and DST data (based on Kheidri et al. 2016).

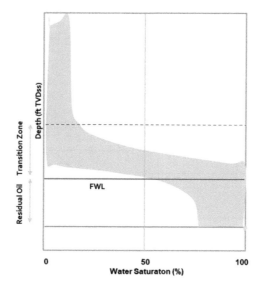

Figure 3.38 Plot of ranges in water saturation versus depth for all wells (after Kheidri et al. 2016).

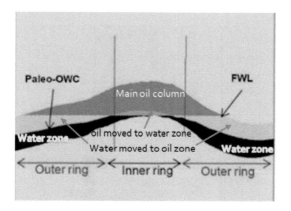

Figure 3.39 Paleo OWC geometry (after Kheidri et al. 2016).

It is worth noting that the paleo OWC in this field seems to have a variable depth and that it also varies in height from the periphery toward the center of the field (see Figure 3.39).

The modeling proposed for the paleo zone by Kheidri et al. (2016) consists of the following steps:

- With respect to porosity and permeability, the diagenetic facies were used as the main driver for modeling residual oil. For the zone below the FWL, S_w logs were upscaled and distributed in 3D using a sequential Gaussian simulation algorithm. Below the contact between the paleo zone and oil/water, S_w was assumed to be 100%.

- A S_w grid for the transition zone between residual oil and the paleo and oil–water contact was obtained by merging the grid from the previous step below the FWL with the grid generated by means of the saturation height functions.

The presence of residual oil below the FWL acts as a baffle impairing injectivity and water mobility. Relative permeability in the presence of trapped oil saturation (including hysteresis) should be used to mimic the baffling behavior.

3.3.4 Tar mat

A *tar mat* consists of highly biodegraded oil sandwiched between aquifer and reservoir. Several investigators have analyzed the geochemistry of the tar mat (e.g. Almansour et al. 2014). A tar mat's typical viscosity is generally 10,000–50,000 cP at reservoir conditions with an asphaltene content between 20–60 wt%. Moor (1984) identified the following four tar mat formation factors:

1. *Water washing*: A portion of light HCs is removed, allowing the asphaltic fraction to be located at the bottom of the oil accumulation.
2. *Gravity segregation*: Heavier HCs are pulled by the resistance toward the bottom, while lighter HCs move upward.
3. *Natural deasphalting*: Natural gases enter from the source rock and rise through the HC column due to buoyancy, resulting in lower solubility and causing the asphaltic fraction to precipitate at the bottom of the reservoir.
4. *Biodegradation*: Meteoric water moves beneath the pooled reservoir and carries bacteria that metabolize the lighter fraction of the crude oil. Reservoir thermal currents distribute this lighter fraction of the rude oil to the oil/water located at the base, where the bacteria are active. As a result, tar mats form at the interface between aquifer and the reservoir.

Tar mat identification in exploration wells is crucial because:

- The tar mat cannot be easily recovered and thus is not to be considered a reserve.
- Tar mats may occur over large areas and display high thickness; they can function as vertical permeability barriers, isolating the oil leg from the aquifer and thus impeding the bottom water drive production mechanism.
- A viscous oil tar mat may mobilize and contaminate the produced oil.

Surveillance is required to characterize the presence of a tar mat. Akkurt et al. (2008) discussed the petrophysical framework for tar mat detection using logging-while-drilling (LWD) measurements. This methodology was implemented in two different carbonate fields in Saudi Arabia, demonstrating that with proper technology and robust interpretation algorithms in place, real-time tar mat identification can be achieved efficiently and accurately. The results further confirmed that existing LWD logging tools are reliable and capable of accurate measurements. Nascimento and Gomes (2004) characterized tar mats in a field using nuclear magnetic resonance (NMR) and conventional logs, and they recommended using the following tools to confirm the presence of tar mats (see Figure 3.40):

78 Integrated Aquifer Characterization and Modeling

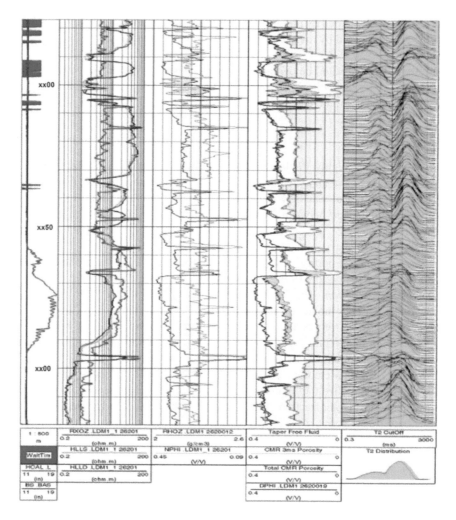

Figure 3.40 Log responses in well A. A tar mat was identified in the interval xx52/xx92 m, from the T2 distribution (track 5), large total NMR porosity deficit (track 4), slight neutron porosity deficit (track 3), resistivity (Rxo) higher than Rt (track 2) and wash out (track 1) (Nascimento and Gomes 2004).

- NMR logs in conjunction with conventional logs can provide accurate identification of tar mat levels.
- Viscosity estimation from NMR logs using empirical correlations can be used for additional confirmation.
- The low hydrogen index in the tar mat zone causes total porosity values estimated by NMR and density logs to be different from those measured in the aquifer and oil zone. Neutron porosity is also somewhat affected by tar mats. Moreover, due to the low mobility of tar mats, resistivity logs show different responses form the oil leg.

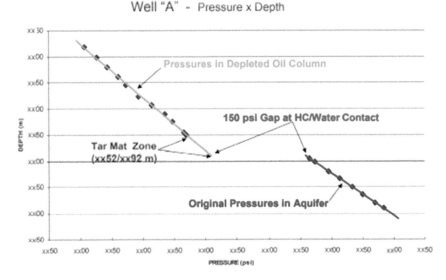

Figure 3.41 Depth versus pressure cross-plot in well A. Pressure distribution shows the evidence of a permeability barrier between the oil column and the aquifer generated by the tar mat at the bottom of the oil column (Nascimento and Gomes 2004).

- The wash out in the tar mat interval is caused by low bitumen mobility. In the oil leg and in the aquifer—where invasion is effective—mud cake forms in the wellbore, keeping the caliper near to the bit size diameter. In the tar mat zone however—where filtrate invasion is more difficult—no mud cake is formed, and the wellbore is enlarged by erosion from mud circulation.
- Original aquifer pressure from wireline pressure data and post-production oil column after depletion can confirm tar mat presence (see Figure 3.41).

Under a high pressure differential across its surface, the tar mat zone can be breached, which may provide the necessary energy for reservoir production. Tar mat thickness and extension knowledge is needed to arrive at the required pressure differential. Alternately, injected solvents can dissolve and make the tar mat mobile. Water injection may be more effective for achieving this provided it is done above the tar mat and at the transition zone.

3.4 ANOMALOUS WATER

The occurrence of water up-dip of a known or assumed OWC or GWC within a reservoir interval could be indicative of:

- perched water trapped during HC migration but still connected to the reservoir
- a separate reservoir segment with a shallower OWC contact
- an isolated compartment (not connected to the reservoir)

80 Integrated Aquifer Characterization and Modeling

Correct identification of anomalous water is important because of its impact on volumetric uncertainty and the risk of structural/stratigraphic complexity it carries due to isolated compartments and/or non-connected reservoir segments with different OWCs that impact water production and water-cut.

3.4.1 Perched water

Perched water is a water volume that occurs as a water-saturated zone separated from the regional aquifer by rock with low S_w. Perched water can occur in groundwater and petroleum fields. Perched water typically occurs in isolated bodies and usually requires a relatively impermeable zone separating it from the main aquifer. A perched-water zone is created when infiltrating water moves vertically through a low permeability layer toward a high permeability layer which acts as local storage (Gaafar et al. 2015). An example of a typical perched water zone is shown schematically in Figure 3.42.[1]

One way of identifying perched water is by means of the MDT data, which in the event of perched water presence may display a separate OWC when compared to the main aquifer gradient, as shown in Figure 3.43.

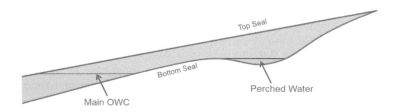

Figure 3.42 Typical perched water zone.

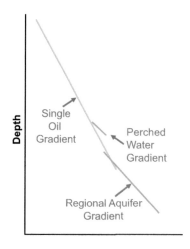

Figure 3.43 MDT data showing perched water contact.

Aquifer description and characterization 81

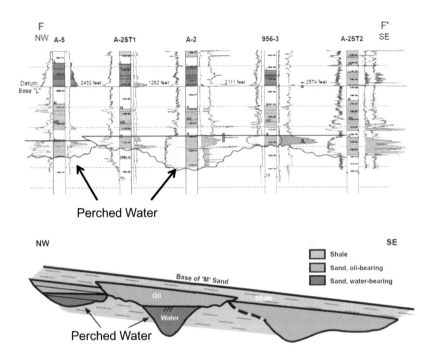

Figure 3.44 Ram Powell Miocene perched water (Kendrick 1999, published with permission).

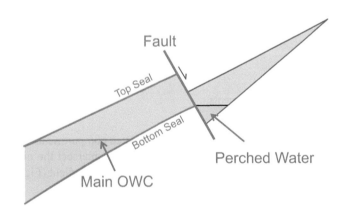

Figure 3.45 Fault-controlled perched water.

A field example of perched water in a Ram Powell Miocene reservoir is shown in Figure 3.44.
One of the factors of perched water formation is faulting, as shown in Figure 3.45 below. A structural saddle is another cause of perched water formation, as shown in Figure 3.46. Water can also accumulate in isolated compartments, as shown in Figure 3.47.

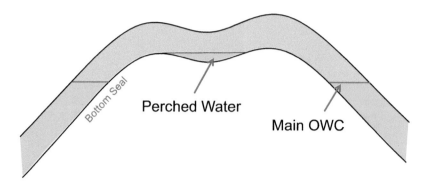

Figure 3.46 Perched water due to a structural saddle.

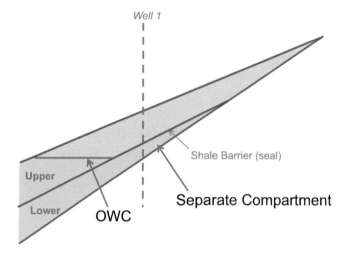

Figure 3.47 Water in isolated compartments.

A key indicator for the presence of water-bearing isolated compartments within a reservoir is when the measured water gradient in a well (a) does not intersect the oil gradient at the apparent well OWC and (b) does not match the regional water gradient (Figure 3.48).

3.4.2 Quick checklist for perched water identification

Perched water is mainly controlled by geological structure and/or changes in facies. Below we provide a checklist for the further characterization of perched water:

- *Pressure data*

    Are water and HCs in equilibrium?

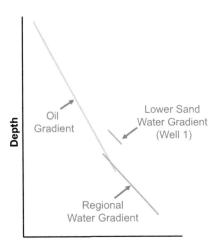

Figure 3.48 MDT plot for water in isolated compartments.

- *Petrophysics*

  HCs are (a) on/in water or (b) separated by a potential shale barrier?
  Does S_w indicate the presence of a transition zone?

- *Water chemistry*

  How do salinity and/or water chemistry compare to the known aquifer water (if penetrated)?

- *Seismic observations*

  Is there evidence of structural or stratigraphic features that could locally trap or isolate water?

- *Geologic explanation*

  What geologic explanation best explains all of the above observations?

3.4.3 Hydrodynamically tilted contacts

As discussed in Chapter 2, under aquifer hydrodynamic conditions, a **water-oil contact (WOC or OWC, which are used interchangeably)** can be developed due to lateral pressure variations. Many examples from around the globe have been described as containing such tilted contacts, some of which are listed in Figure 3.49. Hydrocarbon–water contact (HWC) tilt angle is shown as a gradient in m/km in this figure. Although OWC tilt angles are small, i.e. usually $< 2°$ (left-hand scale of the figure), the impact across an oil field with a diameter of several km can be highly significant (right-hand scale).

The hydrodynamic tilt of an OWC, $\frac{\partial z}{\partial x}$ (where ∂z stands for change in height and ∂x stands for change in x direction), can be expressed by the following simple mathematical equation:

84 Integrated Aquifer Characterization and Modeling

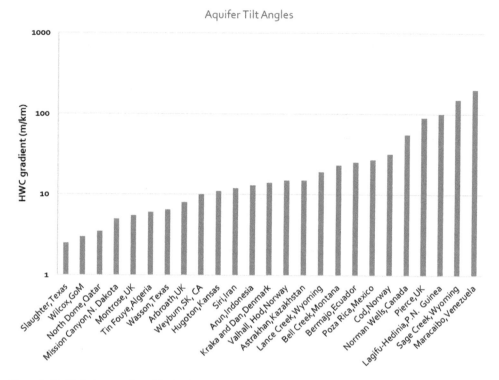

Figure 3.49 Global examples of hydrodynamically tilted WOCs (based on data in Dennis et al. 2000).

$$\frac{\partial z}{\partial x} = -\frac{\frac{\partial p}{\partial x}}{\frac{\partial p}{\partial h_{w-h}}} \tag{3.4}$$

where

$\frac{\partial p}{\partial x}$ is the horizontal component of pressure gradient in the aquifer, w and h are the water and hydrocarbon phases respectively,

$\frac{\partial p}{\partial h_{w-h}}$ is the difference in vertical pressure gradients between the aquifer water and hydrocarbon phases (expressed as pressure per unit column height).

As noted by Dennis et al. (2000), the OWC dip in hydrodynamic equilibrium is not related to capillary pressure. Capillary entry pressure does affect (a) the ability of HCs to migrate through a carrier/reservoir system, and (b) the height of the transition zone once the HCs have come to rest. However, capillary entry pressure has no influence on OWC dip under conditions of hydrodynamic equilibrium.

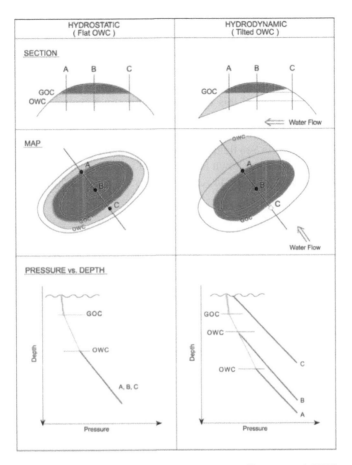

Figure 3.50 Hydrostatic versus hydrodynamic oil–water contacts (Dennis et al. 2005).

Interestingly, even small pressure fluctuations within an aquifer can produce surprisingly large OWC tilts. For example, for a typical aquifer pressure gradient of just 5 psi/km around a medium gravity oilfield, the estimated OWC tilt is roughly 12 m/km. According to Darcy's law, in an aquifer with a permeability of 10 md, such a pressure gradient would cause the groundwater to flow at a rate of 3 cm³ per year (yr), which is a low flow rate compared to that of water injection. However, for a basin with an area spanning 200 km by 100 km expelling water upwards at a flow rate of 0.01 mm³/yr, the resulting 200,000 m³/yr expulsion is comparable with the volumetric flow along an aquifer horizon measuring 100 km by 70 m and with a flow rate of 3 cm³/yr (Dennis et al. 2000).

A comparison of a static and dynamic WOC is shown in Figure 3.50. If a contact is hydrodynamically tilted, the aquifer pressures decrease in the direction of OWC dip, whereas the oil pressure gradient remains constant as shown in this figure. Notice that the latter is like the situation with perched water. However, the potentiometric trends and trap nature distinguish tilted WOC from perched water. As was discussed previously, perched WOCs tend to

86 Integrated Aquifer Characterization and Modeling

be associated with sealing faults and/or interbedded units toward the base of reservoirs whilst hydrodynamic OWCs can occur anywhere within unfaulted, thick permeable beds (Dennis et al. 2005).

Dennis et al. (2000) mentioned the North Sea Valhall/Hod field as an example of a tilted OWC. The interpreted OWC dip averages 1° (15 m/km), indicating a lateral pressure gradient of 7 psi/km. This gradient was found to be consistent with mapped regional aquifer pressures. A pressure gradient of this magnitude acting on a typical chalk aquifer with an average permeability of 1 md could induce hydrodynamic flow at 4 mm^3/year or 4 km^3 every million years.

Tilted OWC in many North Sea fields are shown in Figure 3.51. In 3.51a, 3.51b and 3.51c, the results of 3D simulations are shown for Pierce field, Arbroath field and Blane field. These simulations took thousands of years to reach steady state and match the well pressure data.

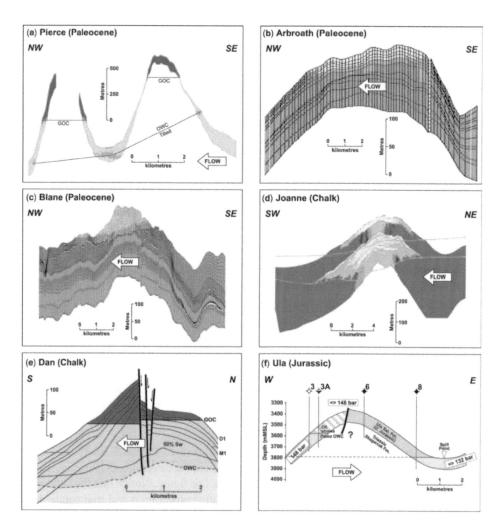

Figure 3.51 North Sea fields with tilted OWCs (Dennis et al. 2005).

Note how in 3.51a, the OWC flattens to the west of the field as it passes into higher permeability sandstone. In 3.51c, the OWC flattens to the left of the section where it passes into an upper layer with higher permeability. In 3.51d, Joanne Field OWC was mapped from well and pressure data, including capillary height curves from horizontal wells. There are two stacked reservoirs (the Ekofisk and Tor formations), both with tilted OWCs. The OWC surfaces are highly irregular as they pass through the heterogeneous aquifer. In 3.51e, Dan Field OWC is shown. Note how it steepens across the central faulted zone. This infers either that the fault is a seal, or that the oil may still be out of equilibrium, leading to lower oil pressures on the left side of the fault. In 3.51f, oil is trapped on the right side of the Ula fault by the positive pressure differential across the fault.

Dennis et al. (2000) also discussed some unusual subsurface configurations and the impact on groundwater flow, as shown in Figure 3.52, which is divided into six sections numbered (a) through (f) from top to bottom. In 3.52a, a typical tilted OWC is shown. In 3.52b, the aquifer dynamic flow can trap the oil below the gas cap. The shape of the contacts below the oil and gas zones is somewhat different due to the density difference between oil and gas. The flow could also flush a shallow trap, as shown in 3.52c. A monocline trap may form if the tilting is aligned with the structure slope, as shown in 3.52d. Aquifer flow may also move some of the oil into an overfilled trap, as illustrated in 3.52e. A monocline trap can become a downslope stratigraphic trap if there is a seal against the oil flow, as shown in 3.52f. It is important to relate unusual well results to hydrodynamic models. Traps which at first appear to be dry, underfilled or even overfilled may instead have a hydrodynamic component.

Finally, for any field which is suspected of being in a hydrodynamic environment, 3D simulation is required prior to developing that field. We agree with the recommendations of Dennis et al. (2005):

> Using the models, it is possible to rationalize hydrodynamic phenomena observed in existing fields and to predict the distribution of oil/gas in undrilled parts of fields and in exploration prospects. This, in turn, can have a marked influence on reserves and, thus,

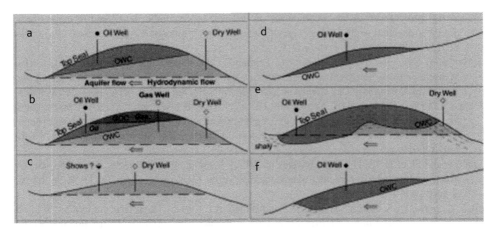

Figure 3.52 Examples of hydrodynamic effects on trapping styles: (a) Tilted OWC, (b) oil and gas separated, (c) shallow trap flushed (d) monocline trap, (e) underfilled and overfilled traps, (f) down-slope stratigraphic trap (after Dennis et al. 2000).

on exploration, appraisal and development strategies. The difference between a conventional oil-field simulation and a hydrodynamic simulation is that the latter is run at very low-pressure differentials and over much longer time periods.

3.5 SUMMARY

We have presented a series of short overviews pertaining to various techniques that are used by the oil and gas industry in identifying and characterizing hydrocarbon accumulations and their associated aquifers. These techniques were grouped by technical "functions" as well as a special section on what we termed "anomalous water." We have discussed that water can be stored in large quantities within regional aquifers, as well as within anomalous zones. These anomalous bodies of water contained in reservoir (permeable) rock can be difficult to predict because they tend to occur in isolated compartments. However, with the right data they can be interpreted/mapped with a reasonable amount of confidence. These interpretations are then integrated into various Earth Models (EM) as will be discussed in Chapter 4.

NOTE

1 We are indebted to Dr. Mick Casey for the concepts in this section and his understanding of anomalous water.

REFERENCES

Abiola, O. and Obasuyi, F.O. (2020). Transition zones analysis using empirical capillary pressure model from well logs and 3D seismic data on 'Stephs' feld, onshore, Niger Delta, Nigeria. *Journal of Petroleum Exploration and Production Technology* 10: 1227–1242. Available at https://doi.org/10.1007/s13202-019-00814-2.

Aikulola, U., Okpobia, O., Okonkwo, A., Yamusa, I, Saragoussi, E., Osabuohien, O., Adegbite, A., Ogunsakin, A., Smith, P. and Hatchell, P. (2020). First time-lapse OBN in Bonga deepwater offshore field. *The Leading Edge* 39 (9): 661–667. http://dx.doi.org/10.1190/tle39090661.1.

Akkurt, R., Seifert, D., Al-Harbi, A., Al-Beaiji, T.M., Kruspe, T., Thern, H. and Kroken, A. (2008). Real-time detection of tar in carbonates using LWD triple combo, NMR and formation tester in highly-deviated wells. Proceedings of 49th Logging Symposium, Austin, Texas, 25–28 May, SPWA-2008-XXX.

Al Mansoori, S.K., Itsekiri, E., Iglauer, S., Pentland, C.H., Bijeljic, B. and Blunt, M.J. (2010). Measurements of non-wetting phase trapping applied to carbon dioxide storage. *International Journal of Greenhouse Gas Control* 4: 283–288. http://dx.doi.org/10.1016/j.ijggc.2009.09.013.

Almansour, A.O., Al-Bazzaz, W.H., Saraswathy, G., Sun, Y., Bai, B. and Flori, R.E. (2014). Investigation of the physical and chemical genesis of deep tar-mat oil and its recovery potential under different temperatures. Proceedings of the SPE Heavy Oil Conference, Alberta, Canada, June 10–12, SPE-170059.

Avseth, P. and Skjei N. (2011). Rock physics modeling of static and dynamic reservoir properties—a heuristic approach for cemented sandstone reservoirs. *The Leading Edge* 30 (1): 90-96. https://doi.org/10.1190/1.3535437.

Ball, V., Tenorio, L., Schiøtt, C., Thomas, M. and Blangy, J.P. (2018). Three-term amplitude-variation-with-offset projections. *GEOPHYSICS* 83 (5): N51–N65. https://library.seg.org/doi/abs/10.1190/geo2017-0763.1.

Berthet, P., Solans, P., Selva, F., Ahn, S.B. and Fournie, J.C. (2015). Production history match of an Angolan deepwater field turbiditic channel using 4D seismic information. European Association of Geoscientists & Engineers. Conference Proceedings. 77th EAGE Conference and Exhibition 2015, Jun 2015, Vol. 2015: 1–5.

Blangy, J.P. (2020). The value and usage of geophysical methods in the estimation of resources and reserves, presented at a post-convention workshop for the 2020 SEG Annual meeting, Houston, October 2020.

Blangy, J.P., Nasser, M. and Maguire, D. (2017). The value of 4D seismic: Has the promise been fulfilled? *The Leading Edge* 36 (5): 407–415. http://dx.doi.org/10.1190/tle36050407.1.

Bretan, P. (2017). Trap analysis: An automated approach for deriving column height predictions in fault-bounded traps. *Petroleum Geoscience* 23 (1): 56–69. http://dx.doi.org/10.1144/10.44pet geo2016-022.

Casey, Mick. (2021). Personal communication.

Chierici, G.L., Ciucci, G.M. and Long, G. (1963). Experimental research on gas saturation behind the water front in gas reservoirs subjected to water drive. Proceedings of the World Petroleum Congress, Frankfurt am Main, sec. II, paper 17, PD6, 483–498.

Connolly, P. 2007. A simple, robust algorithm for seismic net pay estimation. *The Leading Edge* 26 (10): 1233–1360. https://doi.org/10.1190/1.2794386.

Connolly, P., Schurter, G., Davenport, M. and Smith, S. (2002). Estimating net pay for deep water turbidite channels offshore Angola. European Association of Geoscientists & Engineers. Conference Proceedings. 64th EAGE Conference & Exhibition, May 2002, cp-5-00554. https://doi.org/10.3997/2214-4609-pdb.5.G028.

Contreras, A., Gerhardt, A., Spaans, P. and Docherty, M. (2020). Characterization of fluvio-deltaic gas reservoirs through AVA deterministic, stochastic, and wave-equation-based seismic inversion: A case study from the Carnarvon Basin, Western Australia. *The Leading Edge* 39 (2): 82–152.

Crowell, D.C., Dean, G.W. and Loomis, A.G. (1966). Efficiency of gas displacement from a water-drive reservoir. Bureau of Mines. Report of Investigations 6735. United States Bureau of the Interior.

De Gennaro, S., Taylor, B., Bevaart, M., van Bergen, P., Harris, T., Jones, D., Hodzic M. and Watson, J. (2017). A Comprehensivec3D geomechanical model used to deliver safe HPHT wells in the challenging Shearwater Field. Paper presented at the 51st US Rock Mechanics/Geomechanics Symposium, San Francisco, California, American Rock Mechanics Association, 17–735.

Delclaud, J. (1991). Laboratory measurements of the residual gas saturation. Second European Core Analysis Symposium, London, May 20–22, 431–451.

Dennis, H., Baillie, J., Holt, T. and Wessel-Berg, D. (2000). Hydrodynamic activity and tilted oil–water contacts in the North Sea. In: K. Ofstad, J.E. Kittilsen and P. Alexander-Marrack (eds), *Improving the Exploration Process by Learning from the Past*, Norwegian Petroleum Society Special Publications 9, 171–185. Amsterdam: Elsevier.

Dennis, H., Bergmo, P. and Holt, T. 2005. Tilted oil–water contacts: Modelling the effects of aquifer heterogeneity. *Geological Society, London, Petroleum Geology Conference series* 6 (1): 145–158, https://doi.org/10.1144/0060145.

Ding, M. and Kantzas, A. (2001). Measurement of residual gas saturation under ambient conditions. Society of Core Analysts International Symposium, Scotland, SCA 2001-33.

El-Bagoury, M., (2020). Integrated petrophysical study to validate water saturation from well logs in Bahariya Shaley Sand Reservoirs, case study from Abu Gharadig Basin, Egypt. *Journal of Petroleum Exploration and Production Technology*, 10: 3139–3155. https://doi.org/10.1007/s13202-020-00969-3.

Firoozabadi, A., Olsen, G. and van Golf-Racht, T. (1987). Residual gas saturation in waterdrive gas reservoirs. Proceedings of the SPE California Regional Meeting, Ventura, California, April 8–10, SPE-16355.

Flemings, P. (2021). *A Concise Guide to Geopressure: Origin, Prediction, and Applications*. Cambridge: Cambridge University Press.

Gaafar, G., Altunbay, M. and Aziz, S. (2015). Perched water, the concept and its effects on exploration and field development plans in sandstone and carbonate reservoirs. SEG International

Conference and Exhibition, Melbourne, Australia, September 13–16. https://doi.org/10.1190/ice2015-2217766.

Geffen, T.M., Parish, D.R., Haynes, G.W. and Morse, R.A. (1952). Efficiency of gas displacement from porous media by liquid flooding. *Transactions of AIME* 195: 29–38.

Goda, H. and Behrenbruch P. (2011). A universal formulation for the prediction of capillary pressure. Paper presented at the SPE Annual Technical Conference and Exhibition, Denver, Colorado, USA, October 2011, SPE-147078, 1–22. https://doi.org/10.2118/147078-MS.

Hazlett, R.D., Honarpour, M.M., Bulau, J.R. and Vaidya, R.N. (1999). Residual oil saturation dependence on initial water saturation in clean water-wet sandstone. Society of Core Analysts International Symposium, Golden, Colorado, USA, August 1–4, SCA-9912.

Hudec, M.R. and Jackson, M.P.A. (2011). *The Salt Mine: A Digital Atlas of Salt Tectonics*. AAPG Memoir 99. Bureau of Economic Geology Udden Book Series 5. Austin: University of Texas at Austin, and AAPG.

Iglauer, S., Wulling, W., Pentland, C., Mansoori, S. and Blunt, M. (2009). Capillary trapping capacity of rocks and sandpack. Paper presented at the 2009 SPE Europec/EAGE meeting held in Amsterdam, Netherlands, June 8–11, SPE-120960. https://doi.org/10.2118/120960-MS.

Johnston, D.H. and Laugier, B.P. (2012). Resource assessment based on 4D seismic and inversion at Ringhorne Field, Norwegian North Sea. *The Leading Edge* 31 (9): 1042–1048.

Kantzas, A., Ding, M. and Lee, J. (2001). Residual gas saturation revisited. *SPE Reservoir Evaluation & Engineering* 4 (6): 467–476.

Keelan, D.K. (1976). A practical approach to determination of imbibition gas–water relative permeability. *SPE Journal of Petroleum Technology* 28 (2): 199–204.

Kendrick, J.W. (1999). Turbidite reservoir architecture in the Gulf of Mexico: Insights from field development. *AAPG bulletin* 83: (8): 1322.

Kheidri, L.H., Vanhalst, M., Barroso, F., Fadipe, A., Al Hammadi, Y., Al Junaibi, F., Adli, M., Ben Sadok, A., Baslaib, M., Dreno, C., Mel, R., Audigier, P.A., Perroud, S. and Vacheyrout, A. (2016). Modeling of hydrocarbons below free water level in a major oil field in Abu Dhabi UAE and its impacts on dynamic behavior and history matching. Paper presented at the Abu Dhabi International Petroleum Exhibition & Conference, Abu Dhabi, UAE, November 2016, SPE-183507-MS https://doi.org/10.2118/183507-MS.

Kiyashchenko, D., Mateeva, A., Duan, Y., Johnon, D., Pugh, J., Geisslinger, A. and Lopez, J. (2020). Frequent 4D monitoring with DAS 3D VSP in deep water to reveal injected water-sweep dynamics. *The Leading Edge* 39 (7): 471–479. https://doi.org/10.1190/tle39070471.1.

Kleppe, J., Delaplace, P., Lenormand, R., Hamon, G. and Chaput, E. (1997). Representation of capillary pressure hysteresis in reservoir simulation. Proceedings of the SPE Annual Meeting, San Antonio, Texas, October 5–8, SPE-38899, 597–604.

Kralik, J.G., Manak, L.J., Jerauld, G.R. and Spence, A.P. (2000). Effect of trapped gas on relative permeability and residual oil saturation in an oil-wet sandstone. 2000 Annual SPE Meeting in Dallas, TX, Oct. 1–4, SPE-62997. https://doi.org/10.2118/62997-MS.

Kyte, J.R., Stanclift Jr., R.J., Stephan Jr., S.C. and Rapoport, L.A. (1956). Mechanism of water flooding in the presence of free gas. *Petroleum Transactions, AIME* 207: 215–221.

Land, C.S. (1968a). Calculation of imbibition relative permeability for two- and three-phase flow from rock properties. *Society of Petroleum Engineers Journal* 8 (2): 149–156. SPE-1942-PA. https://doi.org/10.2118/1942-PA.

Land, C.S. (1968b). Optimum gas saturation for maximum oil recovery from displacement by water. Paper presented at the annual fall meeting of the Society of Petroleum Engineers of AIME, Houston, TX, USA, Sep. 29, SPE-2216.

Land, C.S. (1971). Comparison of calculated with experimental imbibition relative permeability. *Society of Petroleum Engineers Journal* 11 (4): 419–425. SPE-3360-PA. https://doi.org/10.2118/3360-PA.

Larsen J.A., Trond T. and Haaskjold, G. (2000). Capillary transition zones from a core analysis perspective. Norsk Hydro Research Centre Publication, N-5020 Bergen, Norway. Available from www.researchgate.net/publication/228430681_Capillary_Transition_Zones_from_a_Core_A nalysis_Perspective .

Legatski, M.W., Katz, D.L., Tek, M.R., Gorring, R.L. and Nielsen, R.L. (1964). Displacement of gas from porous media by water. Proceedings of the SPE Annual Fall Meeting, Houston, Texas, October 11–14, SPE-899.

Leverett, M.C. (1941). Capillary behavior in porous solids. *Transactions of the AIME* 142 (1): 152–169. Available from www.onepetro.org/download/journal-paper/SPE-941152-G?id=journal-paper%2FSPE-941152-G.

Ma, T.D. and Youngren, G.K. (1994). Performance of immiscible water-alternating-gas (IWAG) injection at Kuparuk River Unit, North Slope, Alaska. Proceedings of the SPE Annual Meeting, New Orleans, Louisiana, SPE-2860.

McKay, B.A. (1974). Laboratory studies of gas displacement from sandstone reservoirs having strong water drive. *APEA Journal* 14: 189–194.

Miller, B.M. (1982). Application of exploration play-analysis techniques to the assessment of conventional petroleum resources by the USGS. *Journal of Petroleum Technology* 34 (1): 55–64. SPE-9561-PA. https://doi.org/10.2118/9561-PA.

Moor, L.V. (1984). Significance, classification of asphaltic material in petroleum exploration. *Oil & Gas Journal* 82 (41): 109–112.

Mulyadi, H., Amin, R. and Kennaird, T. (2000). Measurement of residual gas saturation in water-driven gas reservoirs: comparison of various core analysis techniques. Proceedings of the SPE International Oil and Gas Conference and Exhibition, Beijing, China, November 7–10, SPE-64710.

Nascimento, J. and Gomes, R. (2004). Tar mats characterization from Nmr and conventional logs, case studies in deepwater reservoirs, offshore Brazil. Paper presented at the SPWLA 45th Annual Logging Symposium, Noordwijk, Netherlands, June 2004. SPWLA-2004-FF.

Nasser, M., Maguire, D., Hansen, H.J. and Schiott, C. (2016). Prestack 3D and 4D seismic inversion for reservoir static and dynamic properties. *The Leading Edge* 35 (5): 415–422. https://doi.org/10.1190/tle35050415.1.

NETL (National Energy Technology Laboratory) (2010). Best practices for geologic storage formation classification: Understanding its importance and impacts on CCS opportunities in the United States. Available from https://netl.doe.gov/sites/default/files/2019-01/BPM_GeologicStorag eClassification.pdf.

Pendrel, J. and Schouten, H. (2020). Facies—The drivers for modern inversions. *The Leading Edge* 39 (2): 102–109. https://doi.org/10.1190/tle39020102.1.

Pickell, J.J., Swanson, B.F. and Hickman, W.B. (1966). Application of air–mercury and oil–air capillary pressure data in the study of pore structure and fluid distribution. *Transactions of AIME* 237: 55–61.

Plug, W.J. (2007). Measurements of capillary pressure and electric permittivity of gas–water systems in porous media at elevated pressures. PhD thesis, Delft University.

Rowan, M.G., Jackson, M.P. and Trudgill, B.D. (1999). Salt-related fault families and fault welds in the northern Gulf of Mexico. *AAPG Bulletin* 83 (9): 1454–1484. https://doi.org/10.1306/E4FD4 1E3-1732-11D7-8645000102C1865D.

Shell (2021). Play based exploration: A guide for AAPG's Imperial Barrel Award Participants. http:// iba.aapg.org/Portals/0/docs/iba/Play_Based_ExplorationGuide.pdf.

Singleton, S. and Keirstead, R. (2011). Calibration of prestack simultaneous impedance inversion using rock physics. *The Leading Edge* 30 (1): 70–78. https://doi.org/10.1190/1.3535435.

Suzanne, K., Hamon, G. and Billiotte, J. (2003). Experimental relationships between residual gas saturation and initial gas saturation in heterogeneous sandstone reservoir. Proceedings of the SPE Annual Technical Conference and Exhibition, Denver, Colorado, USA, October 5–8, SPE-84038.

Swinburn, P., Nayak, P. and van der Weiden, R. (2011). Use of seismic technology in support of reserves booking, Gorgon Field, Australia. Paper presented at the SEG Annual meeting, San Antonio, September 2011, 1129–1133.

Van Gestel, J.P. (2021). Ten years of time-lapse seismic on Atlantis Field. *The Leading Edge* 40 (7): 474–552. https://doi.org/10.1190/tle40070494.1.

Webb, B., Salerno, S., Rizzetto, C., Panizzardi, J., Marchesini, M., Fervari, M., De Draganich, C., Colombi, N. and Calderoni, M. (2020). A time-lapse case study in West Africa: Integrating disciplines for a complete reservoir study and field management. *The Leading Edge* 39 (2): 82–152. https://doi.org/10.1190/tle39020110.1.

Chapter 4

Static models for reservoirs and their aquifers

INTRODUCTION

This chapter addresses the construction of static models that serve as the subsurface framework for dynamic models of fluid flow through time. The types of data analyses and interpretations discussed in Chapters 2 and 3 provide the inputs to these static models. Hopefully, readers are now convinced that the properties of aquifers vary spatially and that it is our job as geoscientists and engineers with O&G insight to capture that variability in 3D models that approximate the Earth in a realistic manner, enabling us to simulate fluid flow.

We can distinguish between three types of geological models that represent various scales of investigation and focus:

(1) basin models at the scale of geological basins or plays
(2) reservoir quality models at the sub-play scale
(3) integrated geomodels at the field scale

4.1 BASIN MODELS

4.1.1 Definition and functionality

A basin model consists of an integrated numerical forward simulation of fluids through geological time and can be performed in either 1D, 2D or 3D. To accomplish this task, basin modeling captures the key geological, physical and geochemical phenomena that occur in the subsurface during the evolution of the geological basin. Key basin modeling steps include simulating the deposition of sediments, their burial, compaction and diagenesis through time, the transfer of heat between sediments and fluids, the evolution of hydrocarbon (HC) source rocks (if present in the basin) through maturation and HC expulsion, and finally the flow dynamics of HC and water. In other words, basin models attempt to capture regional scale fluid flow through geological time: They identify the sources of potential fluid migrations, the mechanisms for fluid movement and expulsion, and they reconstruct these fluid migrations in time and in space.

Basin models have been used in the petroleum industry to aid in assessing exploration potential and geological risk (i.e. the chance of finding HC) in a given area. They help predict, calibrate and assess the effectiveness of the source rock (when present), the timing and type of HC expulsion (oil or gas), the migration routes between the HC kitchen and HC-trapping reservoirs, including the seal capacity required to achieve certain HC column heights. As such, basin models also capture the preservation of petroleum traps and/or their associated hydrocarbons.

Input data requirements for basin models include structural surfaces (or maps) of key intervals from the surface to the source rock (including reservoirs, seals and unconformities) as well as knowledge of the source rock type, regional thermal regime and key formation ages framing the geological history of the area. When available, data from various well control points pertaining to temperatures (used to calibrate heat flow, vitrinite reflectance, etc.), pressures (from MDTs for example), porosities, HC shows and/or fluid samples (i.e. oil API, viscosity and GOR) is used to calibrate the model. Moreover, analyses of headspace gas (in the overburden) and of gas/isotope ratios from logs are important data sources for detecting and understanding the presence of a HC migration front. If major regional faults are known, they are incorporated into the basin model as well. Last but not least, when available, data pertaining to capillary entry pressure is used to characterize capillary seals from the source rock to the surface.

Basin modeling *output calculations* involve three major types of outputs leading to an overall prediction of the petroleum system at the basin scale. Output calculations are used to reconstruct:

- rock compaction processes, which involve the burial history and the evolution of stresses and fluid pressures through geological time
- the thermal history of an area
- HC generation and migration through geological time

Note the timing of the oil and gas windows in Figure 4.1, which shows an output example from a 1D basin model illustrating the burial history for key geological intervals at the site of the northeastern well in the Shams oil field (Shams-NE) in Egypt's Western Desert. Figure 4.2 is from a 3D basin model showing oil migration pathways from various HC kitchens toward potential traps in the subsurface.

Because basin models require a large number of variables as input and predict many types of outputs (temperature and pressure, source rock quality, HC expulsion timing, seal capacity, migration routes for various fluids from pressure sources and sinks), they are most effective when provided with considerable calibration to local data, namely wells and seismic.

4.1.2 Applying physics: solving for pressure and temperature

There are several commercial software packages that perform basin modeling, such as PetroMod, Temis, Trinity, MigMOD, Migris, Novva, TecMod, etc. Basin modeling workflows will continue to be used in the future (Baur et al. 2018) because they are good tools for testing alternate geological scenarios. Basin modeling principles are based on solving for *pressure* and for *temperature* as functions of geological time.

Static models for reservoirs and their aquifers 95

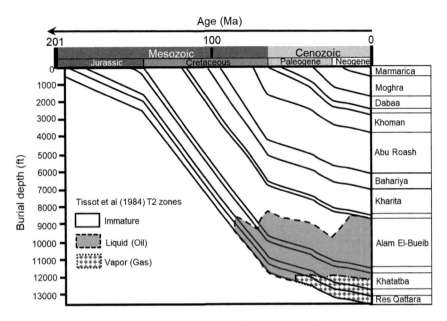

Figure 4.1 1D model showing the burial history as well as the oil and gas window for key geological intervals at the Shams-NE site. (Modified after Shalaby et al. 2008.)

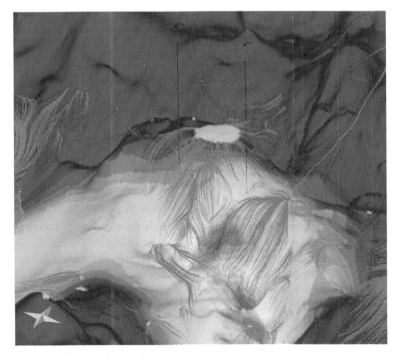

Figure 4.2 An example output of a 3D basin model showing migration pathways from the HC kitchen toward potential traps. (Modified after Bogicevic et.al. 2020, *Search and Discovery*. AAPG©2020, *reprinted by permission of the AAPG whose permission is required for further use.*)

Pressure

We know that fluids flow from areas of high pressure (pressure sources) toward areas of low pressure (pressure sinks). Basin models attempt to capture the evolution of fluid pressures in a given region through geological time. In other words, basin models identify the sources of potential fluid migration and the fluid expulsion and movement mechanisms in order to reconstruct fluid migrations in time and space. Most basin models solve for Darcy's law (as shown below) for two-phase fluid flow by calculating the velocity of each fluid phase (HC and water) within the rock frame. Given the relative permeabilities of fluids present in a rock system, various forces are in play—capillary forces, buoyancy forces, hydrodynamic forces—all of which act within that system:

$$\vec{U}_i = -\overline{\overline{K}} \frac{kr_i}{\mu_i} \left(\vec{\nabla}(P - \rho_w g z) + \vec{\nabla}(P_c) - (\rho_w - \rho_i) g \vec{\nabla}(z) \right) \quad (4.1)$$

$$\uparrow \qquad\qquad \uparrow \qquad\qquad \uparrow$$
viscous term   capillary term   buoyancy term

where

$\vec{U}_i$ is filtration velocity for phase i (m/s)
$\overline{\overline{K}}$ is permeability (m²)
kr_i is relative permeability for phase i
μ_i is dynamic viscosity for phase i (Pa.s)
P is pressure (Pa)
ρ_i is density of phase i (kg/m³)
g is gravitational acceleration (9.81 m/s²)
z is distance (m)
P_c is capillary pressure (Pa)

Unit conversion:

1 Da = 0.987 10^{-12} m²
1 poise = 0.1 Pa.s
1 psi = 0.0001450 Pa
1 lb/ft³ = 0.062428 kg/m³

Typical outputs from pressure calculations display fluid flow vectors at various geological time steps. An example of a calculated output is shown in Figure 4.3. In the figure, note (a) the lateral movement through permeable zones, which are sandwiched between sealing lithologies, and (b) the vertical fluid flow (in the shallow section), which occurs above a top seal (possibly as a result of a top seal having been breached).

Temperature

Basin models also predict heat transfer throughout geological time by solving the heat equation, which describes the balance of energy (expressed as a change in temperature) for

Static models for reservoirs and their aquifers 97

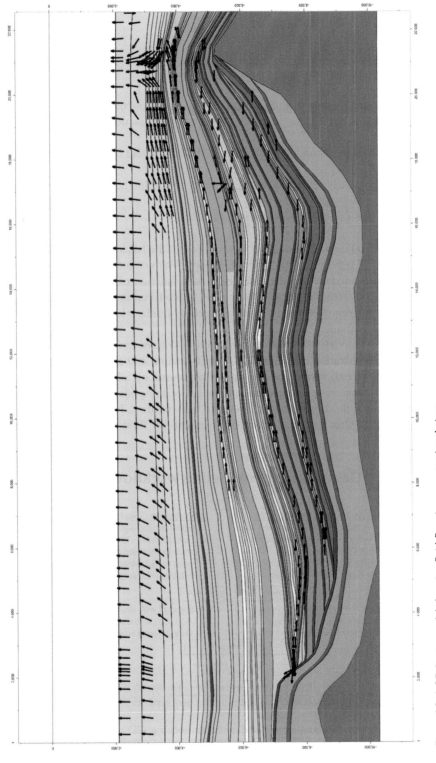

Figure 4.3 A 2D basin model showing fluid flow in cross-sectional view.

heat conducted into and out of a system. There are three basic terms involved—a radiogenic production term, a conduction term, and a convection term:

$$\rho c \frac{\partial T}{\partial t} = \nabla \lambda \nabla T + \rho_p c_p \nabla(\vartheta_p T) + Q_r \qquad (4.2)$$

$$\uparrow \qquad\qquad \uparrow \qquad\qquad \uparrow \qquad\qquad \uparrow$$
temperature　　conduction　　convection　　radiogenic
change　　　　　　　　　　　　　　　　production

where
- λ, ρ, c are bulk thermal conductivity tensor, bulk density and bulk specific heat capacity, respectively;
- ρ_p, c_p, ϑ_p are pore fluid velocity vectors, density and specific heat capacity, respectively;
- Q_r is bulk radioactive heat production.
- All units for equation (4.2) are in metric, SI units

The amount of radiogenic production is a function of the basement type within the upper crust, as well as of sediments and their mineralogy. The conduction term is derived from Fourier's law and utilizes thermal conductivity as a function of mineralogy and porosity. Lastly, the convection term includes two processes, i.e. convection and advection. Convection describes heat being removed from a particular cell by fluids, while advection is heat being supplied to a cell by fluids.

A common basin modeling workflow (see Figure 4.4) consists of importing and analyzing the input data in preparation for calibration, quality control and editing; both the input data and the key (geological) time-steps are further processed by adding subdivisions for age (timing of events) and/or vertical/lateral refinement of the grids; lastly, running the basin model simulations and post-processing complete the workflow. Post-processing consists of extracting results (source rock maturity maps, present-day fluid distribution maps, current

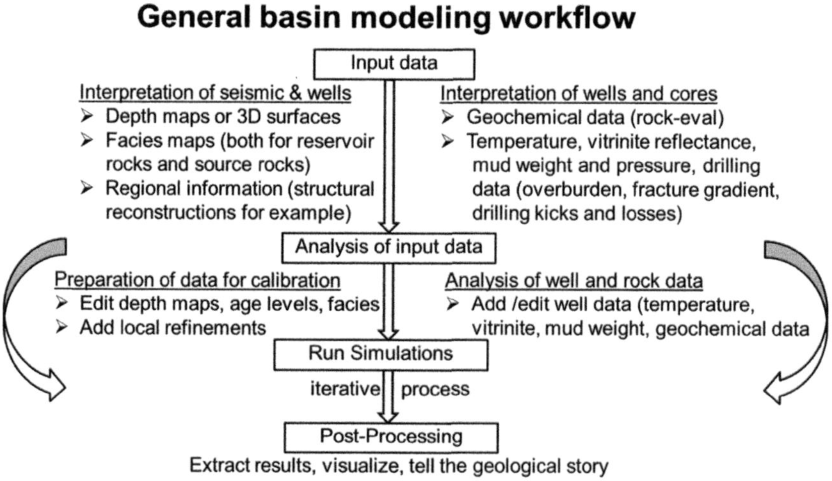

General basin modeling workflow

Input data

Interpretation of seismic & wells
- Depth maps or 3D surfaces
- Facies maps (both for reservoir rocks and source rocks)
- Regional information (structural reconstructions for example)

Interpretation of wells and cores
- Geochemical data (rock-eval)
- Temperature, vitrinite reflectance, mud weight and pressure, drilling data (overburden, fracture gradient, drilling kicks and losses)

Analysis of input data

Preparation of data for calibration
- Edit depth maps, age levels, facies
- Add local refinements

Analysis of well and rock data
- Add /edit well data (temperature, vitrinite, mud weight, geochemical data

Run Simulations

iterative process

Post-Processing
Extract results, visualize, tell the geological story

Figure 4.4 A general workflow commonly used for basin modeling. (2D or 3D).

pressures and temperatures, etc.) and checking their fitness within and against the frame of a coherent and plausible geological story for the area.

4.1.3 Current challenges and limitations in basin modeling

We have discussed how basin models are a great tool for integrating data from multiple disciplines. As such, they are an excellent vehicle for testing alternate geological scenarios in a basin or sub-basin. Current trends in basin modeling focus on increasing the accuracy of simulations of geological processes by making predictive models more quantitative. Three key areas of technology development that integrate basin modeling with other disciplines are structural geology, reservoir engineering and simulating diagenesis:

4.1.3.1 Coupled structural evolution/basin history

When modeling a basin where subsurface deformations have occurred over a given geological duration, it is necessary to reconstruct the movement and structuring of rock layers through time. For example, one type of subsurface deformation is structural uplift, which can be caused by a number of factors, from mega-scale processes such as compressive stresses due to large tectonic forces to local processes such as the unloading/melting of glaciers or the erosion of parts of the overburden. The amount of structural movement in a basin can be significant (on the order of several km) both laterally and vertically, so—depending on geological circumstances—structural restoration can be very important. While a description of its principles and details is beyond the scope of this book, structural restoration is worth mentioning because emerging software technologies are starting to couple basin models with structural and depositional models.

4.1.3.2 Multi-phase compositional fluid flow

Basin models were not initially designed to simulate the relative fluid flow occurring as fluids with various compositions migrate from the source kitchen over the course of geological time. Capturing compositional fluid flow requires both information on those fluids' relative permeabilities and a substantial amount of computational power—similar to that required for modern reservoir engineering simulators. However, current basin modeling research is exploring ways to approximate multi-phase fluid flow through geological time.

4.1.3.3 Diagenesis

Despite the potential support that basin models can bring to reservoir quality predictions, basin modelers are seldom directly involved in performing detailed predictions of reservoir quality. Instead, reservoir quality is addressed either by subject matter experts (SMEs) or through specific sub-processes such as rock physics interpretations of well and seismic data and/or seismic inversion (Chapter 3). Effective stress is defined as the total stress on the rock minus the fluid pressure. In order to predict reservoir porosities, reservoir quality experts use detailed knowledge of rock and fluid compositions, as well as of temperature and effective stress through time. Reservoir quality experts will also use fluid inclusions, fission tracks and paragenesis (knowledge of the chemical precipitation of various constituents like quartz and calcite) to refine and calibrate predictions of reservoir quality at the local scale. However, the quality of these predictions depends on the accurate quantification—using basin models—of the evolution of temperatures and pressures. For this reason, future basin modeling is likely to expand the incorporation and further integration of input from multiple disciplines, such as structural geology, geochemistry, quantitative geophysics, rock quality analysis, rock physics, geomechanics, etc.

In conclusion, 2D and 3D basin models are excellent tools for characterizing hydraulic connectivity at a large scale and for assessing key elements in a basin's petroleum system. They are forward models that utilize key inputs from the types of geological, geophysical and petrophysical analyses described in Chapter 3. As discussed, basin models are used to predict the evolution of pressure and temperature through time at a given location or "cell" in a basin or sub-basin, and to capture fluid movement through time, as shown in Figure 4.3. In addition to pressure and temperature, basin models also predict other geological occurrences through time, such as HC generation/maturation, expulsion and migration; present-day HC phases and fluid quality via PVT properties such as viscosity and GOR; basic diagenesis, vertical effective stress, seal capacity, capillary seal integrity, hydraulic fracturing, biodegradation, biogenic gas generation, etc.

Basin model output can also help assess *rock quality* (RQ) through predictions of rock diagenesis (defined in Section 4.2). For instance, knowledge of an area's temperature and pressure history can be used to estimate how much cementation might occur within the reservoir rock and its associated aquifer from predictions of quartz precipitation via fluid flow through geological time. For this reason, basin modelers work closely with reservoir quality SMEs who provide a more comprehensive and detailed prediction of RQ at the local scale.

4.2 RESERVOIR QUALITY MODELS

Diagenesis is the transformation of sediment into rock. It is an important process that directly impacts the storage and transmissibility properties of reservoirs and seals in HC as well as groundwater systems. The study of diagenesis is extensive and constitutes a specialized subfield of geology (Giles 1997) that utilizes varied data sources ranging from classical petrography to geochemistry, and leverages trend analyses and modern statistics.

Capturing the evolution of a basin includes reconstructing the history of local subsidence of the sediments and their tectonic uplift as well as gaining knowledge about both the fluid volumes and the stages of their movement through time; all this information acts as a catalyst for understanding diagenesis. RQ specialists attempt to quantify the amount of diagenesis in an area by reconstructing the physico-chemical processes occurring at that site and placing them in the context of the geological and hydrological evolution of sedimentary basins. Time is an essential factor of this analysis, so RQ specialists use the concept of thermal stress, which involves both temperature and time, to induce "chemical compaction" on the rock. Circulating pore fluids, which often tend to be oversaturated with silica and/or carbonate and to react with the host mineralogy, eventually attain chemical equilibrium with the host rock they are traversing. Under low temperature conditions typically occurring in shallow sections of the subsurface, diagenesis is mostly driven by chemical kinetics, while under deeper and higher temperature conditions, diagenesis obeys the principles of thermodynamic equilibrium. For details of these chemical and thermodynamic changes see Giles (1997).

Generally speaking, we expect RQ to degrade with depth of burial. This degradation is due to the combination of increased temperature and pressure with depth, which in turn leads to porosity loss with depth. While porosity loss is predictable, it occurs at highly variable rates dependent on a number of variables that are a function of local conditions (i.e. rate of burial, amount of heat flow from the basement, in situ effective stress, availability of various minerals from local lithologies to be either dissolved or transported and then redistributed, etc.). It is the role of the RQ specialist to understand the controls on rock quality for a particular area by accounting for changes in parameters controlling heat, fluid and mass transport through time.

Reservoir quality predictions are commonly used during exploration for prospecting risk and for making decisions on the acquisition of potential new acreage. They are also used throughout development and production for assessing RQ in both oil/gas reservoirs and their aquifers. RQ predictions estimate the effectiveness of key recovery processes and allow for the planning of enhanced recovery if needed (see Chapter 7).

One example of a software system widely used in O&G to analyze controls on sandstone reservoir quality in analog core samples is Touchstone™, initially developed by Geocosm through the Rock Quality Consortium for Quantitative Prediction of Sandstone Reservoir Quality (RQC). Using output information from basin models (such as temperature and pressure), combined with detailed petrographic data and diagenetic models, this software deploys forward models to predict key parameters—away from well control—such as the amount of compaction, quartz cementation, porosity and permeability for sandstones.

Key concepts are presented in the following section without exhaustive use of equations. For readers who want to access more in-depth theoretical analyses, a good reference for equations representing the physics and mathematics (entailing thermodynamics, kinetics and the physics of heat and fluid flow) of diagenesis can be found in Giles (1997).

Rock properties are essential for evaluating potential storage and fluid flow in reservoirs, as well as for understanding the seal capacity of top seals (capillary seals). Two key properties that control fluid storage and flow are ***porosity*** and ***permeability***.

4.2.1 Porosity models

There are a number of established models that predict the behavior of porosity in clastic sediments as a function of depth. As we discussed in the previous section, present-day depth does not have a direct control on porosity, while thermal stress and effective stress do. Nevertheless, porosity and depth are inversely correlated—keeping in mind that correlation is not causation. Two geological processes are in play that concern subsurface depth: First, for shallow depths below the surface, porosity is systematically altered by compaction processes, wherein sediments undergo dewatering as they become buried. Second, as the depth of burial increases so does temperature, and for depths beyond approximately 2 km—depending on the temperature gradient in the area—processes involving mineral cementation and/or alteration begin to occur, further altering porosity.

If water cannot escape normally during the compaction process (occurring at shallow depths), fluid pressure increases and sediments retain a higher porosity than the normal compaction trend; in other words, sediments are under-compacted. Conversely, if compaction is very efficient locally, sediments lose their porosity faster than the compaction curve for the area, creating a pressure regression. This concept of *compaction equilibrium* can be used to predict the porosity of permeable sediments based on their connectivity as a function of effective stress. Most published compaction curves illustrate porosity as an exponential function of depth or effective stress. They start by characterizing the simplest compaction phenomena for quartz-rich sands (Gluyas and Cade 1997), and then include refinements that incorporate mineralogy, i.e. for clay content and/or lithics (Pittman and Larese 1991). Modern compaction models include the effect of total effective stress by incorporating horizontal stresses as well as vertical stress in the formulation for porosity (Flemings 2021).

Reservoir quality prediction also entails understanding thermal stresses through time. Reaction rates increase as temperature rises with depth, and thermal reactions can become predominant. A number of thermal reactions have been described in the literature that characterize the interaction between deep fluid movement and deep subsurface sediments (Giles 1997), such as quartz and/or carbonate cementation, feldspar dissolution, illitization and clay transformation, dolomitization of carbonates, etc. Note that porosity preservation can also be facilitated by thermal reactions. For example, Ramm et al. (1997) show that micro-quartz grain-coating led to porosity preservation below what would normally have been considered economic basement (tight rock). The authors describe how microcrystalline quartz retarded chemical reactions and thus inhibited quartz precipitation and dissolution into framework grains.

There are a number of reservoir quality predictions that formulate porosity as a function of depth for carbonate environments. For example, Brown (1997) argues that mineralogy and shale content are key controlling factors for the rate of porosity loss. He also shows that dolomitic intervals in the Williston basin exhibit better "selective" porosity preservation than limestones. Carbonate leaching and/or cementation is sometimes associated with and constrained within depositional facies (Major and Holtz 1997), and the depositional facies controls the development of specific pore types (Mountjoy and Marquez 1997).

RQ specialists understand why a particular reservoir might be tight. Global diagenesis patterns are described in terms of facies (Primmer et al. 1997). There are a variety of geological causes that can affect reservoir porosity, as discussed by Tobin (1997) and summarized in Table 4.1 (adapted from Tobin 1997).

To summarize, RQ experts reconstruct the unfolding of compaction processes (via effective stress computations) and chemical processes (via thermal stress computations) through geological time in order to predict the porosity of sediments. They calibrate their models with fluid inclusions, fission tracks and other techniques so that results from paragenesis can be iteratively integrated with basin modeling to improve porosity predictions (Schoener et al. 2008). Similar to basin modeling, we expect that porosity prediction practices will garner additional benefits and insights in the future by integrating RQ workflows with rock physics through the application of rock physics templates (RPTs), which can help reconstruct the burial history of sediments (Avseth and Skjei 2011) and/or differentiate cemented zones from less cemented zones (Avseth and Lehocki 2016). Further porosity prediction improvements may arise through the quantitative inversion for rock properties from seismic (see Chapter 3) and through our improved understanding/quantification of the physico-chemical transformations attending geological processes such as diagenesis.

Table 4.1 Key geological factors impacting porosity in the subsurface and main technologies used in porosity prediction

Geological control	Impact on porosity	Technology
Ancient destuctive diagenesis (including sediment infill, mechanical and chemical compaction, cementation, recrystallization)	Reduction	Sedimentology, rock quality analysis, rock physics
Ancient constuctive diagenesis (including dissolution and fracturing)	Increase	Rock quality analysis, rock physics, geomechanics
Other ancient diagenesis (including mineralogical replacement, authigenic clay growth, brecciation, tectonic deformation)	Can either increase or reduce porosity	Rock quality analysis, core analysis, structural analysis
Environment of deposition	Controls pre-diagenesis porosity	Sedimentology, geophysics, seismic interpretation
Framework composition and/or provenance, rock fabric, pore geometry	Possible control on diagenesis (post- depositional)	Sedimentology, petrography, core analysis, luminescence, fluorescence, rock quality analysis
Paleoclimate	Impacts EOD, weathering, karstification	Sedimentology, facies analysis
Depth of burial	Correlated to porosity loss	Compaction curve analysis, seismic interpretation
Pressuring and overpressuring	Early overpressure can enhance porosity	Basin modeling, seismic interpretation, rock physics, geomechanics
Thermal maturity	Related to porosity loss	Basin modeling, rock quality analysis (diagenesis), rock physics
Erosional events and/or unconformities	Can either increase or reduce porosity	Structural analysis, seismic interpretation
Pore fluid migration (water phase)	Enhances cementation and/or dissolution	Basin modeling
Pore fluid migration (oil phase)	Inhibits cementation and/or dissolution	Basin modeling thermal maturity, fluid inclusion thermometry
Associated rock strata—seal	Affects pore fluid entrapment	Sedimentology of shales, seal analysis, geomechanics
Associated rock strata—source rock	Controls the type of migrating fluid	Sedimentology of organic shales, organic geochemistry

4.2.2 Permeability from porosity

In Section 4.2.1 we discussed how sediment porosity is affected by diagenetic processes. Clearly, it is important to understand when the diagenetic changes occurred and to what extent they have transformed the rock's potential storage space. While the porosity of a rock impacts its fluid storage potential, it is the rock's permeability that determines the extent to which fluids can flow through it. Determining permeability is particularly challenging because this parameter depends on a variety of subsurface factors and therefore exhibits a certain degree of heterogeneity as well as directional preference (anisotropy).

Accurate prediction of the spatial distribution of permeability in the subsurface is key to building credible subsurface models. Traditional permeability prediction workflows employ a porosity-to-permeability transformation typically applied by facies class. These are empirical methods that can be very successful at predicting permeability, but they require sufficient calibration to local data, which is not always available. Alternative approaches to permeability prediction (similar to those employed for porosity) attempt to reconstruct the geological processes of burial and cementation, but concentrate on predicting the pore structure of the rock. Important controls on such workflows include mineralogy, sorting, grain size and the amount of ductile grains (Evans et al. 1997).

Permeability is highly dependent on the distribution of pore throat sizes, which are often directly related to the extent of diagenetic transformation. Ehrlich et al. (1984) concluded based on thin sections from rock samples that matrix permeability (which excludes the contribution of fractures) in clastic sediments is controlled by the depositional fabric and by diagenesis, which incorporates geochemical changes in the rock. In order to characterize subtle changes in permeability, the authors defined a number of pore types or categories based on pore throat size distributions. Cementation commonly occurs at depths where temperatures have risen substantially and is due to the circulation of fluids super-saturated with carbonate or quartz. Cementation tends to isolate certain pore throats to the point of partially blocking the circulation of fluids and thus significantly reducing permeability. RQ experts work closely with geologists, petrophysicists and reservoir engineers to assess potential microscopic displacement efficiency, as well as irreducible water saturation and residual HC trapping. The last two parameters control the end-points of relative permeability curves for multi-phase flow (HC and water).

Multiple prior attempts to assess and predict permeability in carbonates control for permeability mainly through diagenetic processes involving dolomitization and leaching or dissolution (Mountjoy and Marquez 1997). Permeability preservation or destruction is often greater in carbonate than in clastic environments because carbonates are more reactive during chemical processes involving dolomitization and dissolution. Nevertheless, Mountjoy and Marquez note that it is often the depositional sub-facies that controls the development of specific carbonate pore types. The authors acknowledge that predicting porosity–permeability relationships in carbonates is notoriously difficult, but they conclude that the development of secondary (diagenetic) porosity in deeply buried dolomites from Alberta is controlled mainly by the distribution of primary porosity, which in turn is impacted by the original depositional facies. Taking the description of the rock fabric one step further, Mur and Vernik (2021) implement a quantitative rock physics approach in their analysis of the factors impacting pore shape and are thus able to predict porosities for a wide range of carbonate types around the world.

Understanding the rock fabric and the diagenetic history of a rock does not make the task of predicting its permeability an easy endeavor. Permeability prediction techniques based on pore type distributions can be augmented with the use of emerging digital rock physics (DRP) workflows that digitize representative rock samples in 3D. DRP workflows automatically classify the rock fabric, grain size distributions and pores using modern computational algorithms whose outputs are then used to reconstruct permeabilities. Emerging permeability prediction trends rely to a greater extent than traditional, established practices on the integration of knowledge from multiple disciplines (rock quality/diagenesis, basin modeling, physics, chemistry) alongside developing digital rock analyses (Byrnes et al. 2018) and artificial intelligence algorithms.

4.3 GEOMODELING: FIELD SCALE INTEGRATION

Geomodels integrate available data from multiple sources within a common digital geological framework that incorporates several different frameworks:

- a structural framework that addresses the size and shape of the overall container of fluids
- a fault framework that addresses potential structural boundaries within the model
- a stratigraphic framework that treats key expected stratigraphic units and their lateral/vertical relationships, pre-development fluid contacts and initial reservoir pressures as established prior to HC extraction

The same principles applied to characterize static HC models can be applied to aquifer models, keeping in mind that only one fluid phase (water) is being modeled. When two or more fluids are present (water and oil for example, or two different water sources), the two phases can interact. The details of the interface between a HC leg and a water leg sometimes indicate a fluid transition zone and are captured within geomodels. An example of a transition zone analysis is covered in Chapter 3 in the section on petrophysical information.

4.3.1 A typical geomodeling workflow

Modern geocellular models are a digital representation of the static state of the subsurface. They capture the structural and stratigraphic variability in reservoir geology and non-reservoir facies—as well as the accompanying distribution of rock and fluid properties in 3D—into a common Earth model. Faisov et al. (2019) discuss ways to increase the quality of 3D geomodels.

The main differences between basin models and field scale geomodels are model resolution and the types of reservoir properties that are populated. One key function of geomodels is serving as building blocks for dynamic reservoir simulations. As such, geomodels are likely to incorporate detailed core-based observations and measurements used to calibrate log data and to estimate rock properties at a fine vertical scale. These property estimates may then be grouped or averaged by facies type. Some examples of core data use are shown in Table 4.2, where RSWC stands for rotary sidewall core and SCAL stands for special core analysis, as opposed to routine core analysis (RCA).

Table 4.2 A non-exhaustive list of core-based observations for geomodeling input

Type of core analysis	Use	Geomodeling application
Continuous measurements (whole core)	Gamma, photography, CT scanning	Indicative sedimentology, mineralogy
Continuous measurements (whole core)	Profile permeability	Indicative permeability
Continuous measurements (whole core)	Hyperspectral imaging	Mineral composition
X-ray diffraction (XRD)	Sedimentology, mineralogy	Sedimentolology (facies identification)
X-ray fluorescence (XRF)	Sedimentology, mineralogy, fluids	Sedimentolology (facies identification)
Chemostratigraphy	Chemical analysis	Sedimentolology (facies identification)
Biostratigraphy	Age dating	Sedimentolology (facies identification)
This section analysis/point count	Detailed mineralogy	Sedimentolology (facies identification)
Core description	Detailed grained description, associations	
Core description	Depositional environments, facies Sedimentologic description Scanning electron microscopy-SEM	Stratigraphic framework
Core description	Fracture description	Structural framework
Thin section analysis	Assessment of diagenesis (porosity, cement, grain or clay alteration)	Rock properties (rock quality analysis)
Integrated core studies	Facies classification Property determination	Stratigraphic framework
RSWC; Description only	Mineralogy Relative porosity	Rock properties (mostly descriptive)
SCAL; measurements	Effective porosity versus total porosity (primary and micro-porosity)	Rock properties (rock quality analysis)
SCAL; measurements	Grain density, bulk density Particle size analysis (laser grain size, sieve)	Rock properties (rock quality analysis)
SCAL; measurements	Continuous scratch test	Calibration of log-derived mechanical properties
SCAL; measurements (preserved core)	Salinity analysis Fluid extraction Fluid saturation determination	Fluid property analysis (residual fluids)
SCAL; measurements (preserved core)	Salinity analysis Fluid saturation determination	Fluid property analysis (residual fluids)
SCAL; dynamic measurements	Absolute permeability (air) Wettability Relative permeability (multi-phase)	Rock and fluid properties (dynamic)

Building a geomodel is a major undertaking that takes several months to complete. A typical geomodeling workflow (see example in Figure 4.5) serves as the building block for a future reservoir simulation. There are two important phases to a geomodeling workflow:

4.3.1.1 Phase 1: building the structural and stratigraphic framework

This step is where the "architecture" or geometry of the geomodel is built, where inputs are inherited and combined from interpretation activities discussed in Chapter 3. Depth

Static models for reservoirs and their aquifers 107

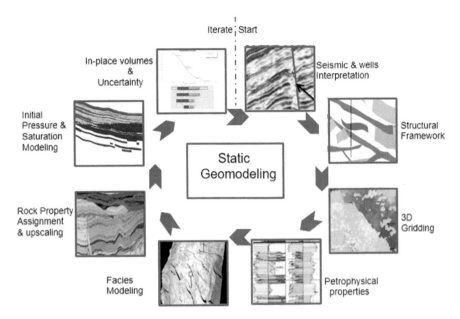

Figure 4.5 A typical geomodeling workflow estimating in-place volumetrics.

interpretations of seismic horizons and well tops are gridded, minimizing any residual depth differences between wells and seismic using criteria based on the distance to the wellbores. The gridding process may involve some local smoothing (for example near mapped discontinuities) as well as a certain amount of lateral averaging. After gridding multiple horizons, a horizon-based framework is constructed in order to estimate the gross thicknesses of key geological packages. Mapping a base seal horizon is not a requirement if the trap is a structural closure, but it is needed for stratigraphic traps. Figure 4.6 displays an example with approximately 12–15 zones in a geological section expanding from west to east across a number of prominent normal faults.

A structural framework consists of polygons assembled from the faults identified from seismic and wells and is constructed at the same time as the gridding of the horizons. Fault polygons included in the structural framework are labeled (a) high-confidence (all seismically observed faults with distinct offset or faults encountered in wells), (b) mid-confidence (where the seismic indicates likely faults that may or may not display an offset) and (c) low-confidence (where there is the possibility of a fault without any visible offset). Note that the likelihood of wells intersecting a fault is relatively low, especially when faults are steep, so it is fair to say that vertical wells are likely to "undersample" the fault population in the subsurface (Figure 4.7).

The resulting structural grid often takes the form of a "pilar grid" in which segmentation is the product of fault connections. While there are various types of gridding methods, a discussion of which to use in what context is beyond the scope of this book. For our purposes here it suffices to note that it is important to consider the criteria for the grid design (horizontal and vertical dimensions) as well as the grid's orientation because the vertical grid size defines the finest layering scheme consistent with stratigraphic zonations and depositional

108 Integrated Aquifer Characterization and Modeling

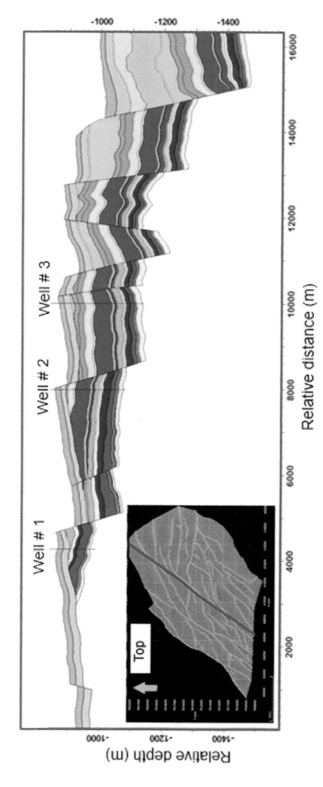

Figure 4.6 Incorporation of fault interpretations from seismic and wells used to form zones.

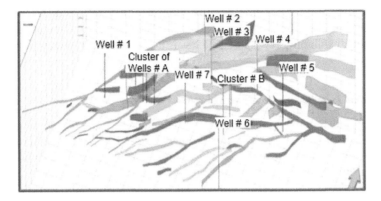

Figure 4.7 Incorporation of fault interpretations from seismic and wells.

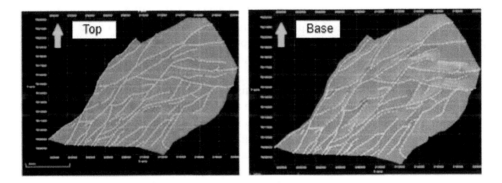

Figure 4.8 Establishing fault grids for each zone: top and bottom.

facies' interpretation scenarios, while the horizontal grid size drives the model's lateral resolution limits. The pilar gridding process is performed for each horizon (Figure 4.8). Because typical geomodels tend to be very large, modelers also account for computational constraints for upcoming simulations. Typical grid cell sizes are on the order of 50–100 m (150–300 ft) on both the x and y axes, and 0.5–1 m (2–3 ft) on the z (depth) axis, often leading to models containing up to 10 MM cells.

At this point in the geomodeling process, it is customary to establish a depth uncertainty map that captures areas where the depth structure may be steeper than portrayed or where the quality of the seismic data used for mapping is poor. It is not unusual for seismic depth maps to carry a depth uncertainty of approximately 1% (depth) away from wells, and sometimes higher in areas of poor data.

Pilar grid geometry is often grouped into model regions. Model segmentation may take the form of a geographic (e.g. NW sector) or geologic basis in some cases. For instance, if faults are believed to be acting as boundaries, modelers may suggest a grouping representing possible regions of segmentation with different pressure compartments. Once the main segments are established, (static) pressure data and fluid contacts—when known—serve to reinforce the geologic model. Large pressure differences between areas lead to compartments with isolated volumes and/or different fluid contacts (OWC or GWC). Like any data, pressure data is of variable quality, exhibiting low confidence levels in low-mobility areas or if measurements are scattered.

4.3.1.2 Phase 2: assigning layer properties

The next step in the modeling process is the assignment of facies classes as inherited from iterative geological and petrophysical workflows. Geological facies classify rock types as key units of related geological bodies with similar rock properties. Common approaches involve deterministic facies estimations, as well as statistical methods (i.e. stochastic sequential simulations or Bayesian assignments). Away from wells, workflows for facies determination leverage—when available—input from seismic rock physics-based inversion/classification methods (see Chapter 3). Data-driven seismic inversion methods tend to strongly reflect the physical constraints contained in the variability of the seismic facies, while statistical or simulation methods reproduce data statistics, allowing for multiple equiprobable facies distributions (Chapter 3, Section 3.2.5). At this stage of the geomodeling process, the most likely stratigraphic framework is revised and refined to ensure that local detail from core, log (chemostrat, image logs, log analyses) and seismic is fully integrated and incorporated.

The geomodel's final facies assignment aims to represent as much geological realism as possible. The example in Figure 4.9 shows the facies assignment for the same section as the one shown in Figure 4.6. The final facies assignment was applied to a grid design increment of 100×100×1.5 without the need for upscaling.

Petrophysical interpretation serves as the basis for assigning layer rock properties on a sample-by-sample basis. Key parameters evaluated at typical log sampling intervals of 15 cm (0.5 ft) include clay content or shale volume, net-to-gross (NtG), porosity and permeability. As discussed earlier in this chapter, permeability is estimated using statistics that reproduce joint experimental distributions of porosity–permeability. Fluid saturation (in this case, S_w) is of particular importance for the energy industry.

Rock properties calculated at a fine sampling interval are often averaged by geological facies and/or potential flow units in a process that upscales from the log sampling scale to the scale of the geomodel's grid. Some properties upscale volumetrically (like total porosity), while others do not. For example, directional properties like permeability or seismic velocity need to be upscaled (and averaged within the upscaled interval) according to the physics of equivalent media. Upscaling is the subject of specialized analysis and is beyond the scope of this book. The upscaling process may rely on physics and/or on the use of geostatistics, stochastic methods or a Bayesian framework. For more on the use of statistical methods see Doyen (2007), Isaak and Srivastava (1990) and Saad et al. (2019).

An example of vertical layering defined from logs and pressure in a well is shown in Figure 4.10. Curves from left to right include shale volume, S_w, gamma ray, resistivity, MDT pressure points, large units or zones defined in the structural framework and model layering according to facies. Note the presence of an aquifer in the lower units.

Static models for reservoirs and their aquifers 111

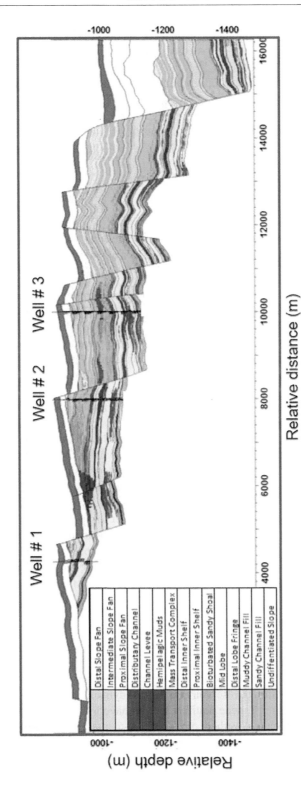

Figure 4.9 Facies assignment within the geomodel. (Same section view as Figure 4.6.)

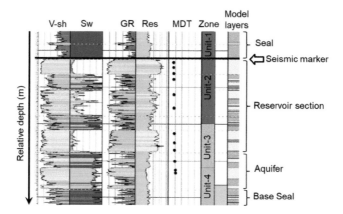

Figure 4.10 Geomodel property determination and facies assignment.

At this stage in the model construction workflow, the geomodel is sense-checked in 3D to ensure that it represents the geology fairly and captures sufficient heterogeneity to conduct a flow simulation. Fluid contacts are revisited if needed, and fluid analyses (including HC composition and water salinity) are incorporated in the pressure interpretations in order to establish a preliminary understanding of possible connectivity pathways. An example of a 3D visualization of the geomodel is shown in Figure 4.11. The model includes a distribution of each of the static properties discussed in the text and averaged at the scale of the grid.

A dynamic model, i.e. a fluid flow model based on the numerical grid derived from a static model, incorporates a *field development plan* (FDP) with a series of associated wells. The difference between a static and a dynamic model lies in the additional dynamic uncertainties pertaining to (a) the potential connectivity and barriers and/or baffles between flow units, as well as (b) the location, number and type of wells. Determining the fluid transmissivity of faults (and whether they act as "permeable membranes," seals or baffles) is particularly difficult without access to production (dynamic) data, especially given the fact that fault transmissivity can evolve (sometimes increasing) with reservoir pressure depletion. Figure 4.12 shows a schema of the main steps involved in the construction of an integrated static-to-dynamic model. Aside from the static model (Figure 4.5), note the addition of an FDP (with the number, location and type of wells) and of production pressure and/or saturation history matching, production forecasting and benchmarking to producing analogs.

4.3.2 Aquifer-specific geomodeling

To be representative of the subsurface, reservoir and aquifer models must capture potential variations in parameters likely to impact connectivity and fluid flow. These parameters—which include fault transmissibility and the extent, properties and relative permeability of aquifers—are adjusted through reservoir simulation (dynamic models).

As discussed in Chapter 3, deep aquifers tend to have high acoustic impedance and low seismic reflectivity, which makes them more difficult to map than HC reservoirs. As a result, mapping the extent of deep aquifers is more difficult than mapping that of shallow aquifers (which are likely low impedance), and their seismic interpretation is also likely to be more

Static models for reservoirs and their aquifers 113

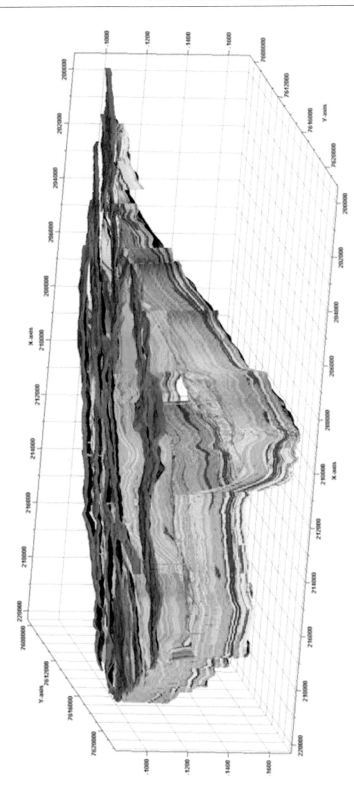

Figure 4.11 A 3D visualization of the geomodel.

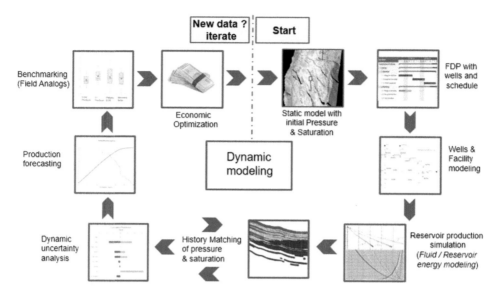

Figure 4.12 The relationship between a static model and a dynamic model.

time consuming. Several aquifers may be present (with different water properties)—some may be perched or isolated, while others may be in communication. Furthermore, aquifer quality may degrade in a more complex manner than HC reservoir quality. Applying a simple linear porosity-permeability-depth trend projected from crestal wells may not be sufficient to capture the degradation of aquifer properties in a given area. Moreover, any communication barriers/baffles (such as faults) present in the aquifer should be mapped. Lastly, if communication between injectors and producers is suspected to be present, potential pathways between the completed production intervals should be identified.

4.3.3 Preliminary volume calculations and volumetric uncertainty assessment

4.3.3.1 Basics of in situ petroleum volumetrics (resources)

In-place oil volumetrics (in O&G field units of STB) are calculated as follows:

$$OOIP = 7758 \, Ah \, NtG \, \varnothing \frac{(1-S_{wirr})}{B_o} \qquad (4.3)$$

where
 A is average area (acres)
 h is average reservoir thickness (ft)
 ϕ is porosity (fraction)
 $S_{w\,irr}$ is irreducible water saturation (%)
 B_o is oil formation volume factor (reservoir bbl or STB).

The number 7758 is a conversion from acre-ft to bbl. For metric units, this factor is removed.

In-place gas volumetrics (in O&G field units of bcf of gas) are calculated as follows:

$$OGIP = 43560\, A h\, NtG\, \varnothing \frac{(1-S_{wirr})}{B_g} \qquad (4.4)$$

where
B_g is gas formation volume factor (in reservoir ft^3/SCF).
The number 43560 is a conversion from acre-ft to ft^3.

For the aquifer, the *OWIP* volume (in O&G field units of STB) is the same as OOIP in Equation 4.3, except B_w is used instead of B_o and $S_{w\,irr}$ is 0.

The geomodel allows for the systematic calculation of volumes within each individual segment and their aggregation into zones/regions or at the total field scale. Due to the uncertainty associated with their exact estimation, each of the input variables exhibits a possible range, generating uncertainty for the output volumes. A common practice for addressing this issue is to specify output volumes in a ***probabilistic manner***, either in the form of a distribution of potential volumes or as a cumulative probability distribution function.

Figure 4.13 shows two types of displays for statistical distributions characterizing the in-place volumes of an oil field (the output of Equations 4.3 and 4.4). The diagram on the top left is a probability distribution function, while the diagram on the top right is an exceedance probability function (where exceedance is defined as 1 minus the cumulative distribution). It is common practice to use the exceedance curve, where the P90 point corresponds to a 90% chance of encountering 680 MMBO, the P50 (most likely case or reference case) is a 50% chance of encountering 968 MMBO and the P10 has a 10% chance of encountering 1338 MMBO. The probabilistic estimation of in-place volumes is often accompanied by a discussion of the key parameters impacting those volumes. This is shown in the tornado chart at the bottom of Figure 4.13, where we can see that for this particular oil field, the trap area uncertainty that is associated with faulting has the largest impact on volumes, followed by uncertainties in saturation, porosity, thickness and NtG.

4.3.3.2 Basics of recoverable petroleum volumetrics (reserves and resources)

In-place resources have no assigned value until they are associated with a FDP, at which point recoverable volumes are treated through an estimated ***recovery factor***. The FDP characterizes the number of wells, schedule and cost associated with putting a field in production, and it drives the recovery factor. Once an FDP is approved through a field development decision (FID), the operator takes on a commitment to the government to proceed with development. For economic decisions, operators and governments usually require a minimum of three ***deterministic development cases***, each labeled as (a) most likely (P50), (b) low side case (P90) and (c) high side case (P10) for recoverable resources. Sources of uncertainty include data quality, known uncertainties, known unknowns and unknown unknowns. It is a good idea for uncertainties to be captured under realistic geologic as well as several engineering scenarios.

116 Integrated Aquifer Characterization and Modeling

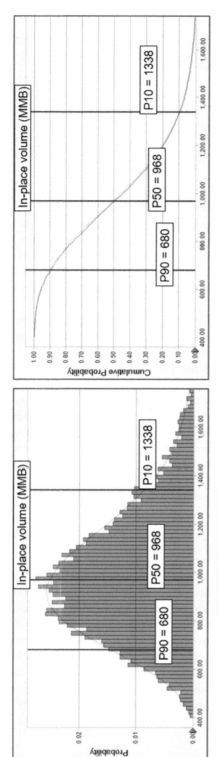
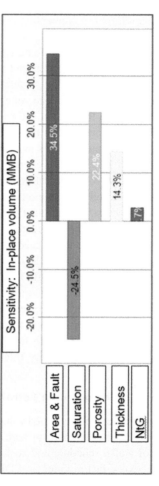

Figure 4.13 Probabilistic determination of static (in-place) volumes and uncertainty analysis.

When determining value or commercial viability, the important variables are production rates and recoverable volumes, cost (number of wells and infrastructure) and schedule. The distinction between reserves and resources is governed by strict rules detailed in the Petroleum Resources Management System (PRMS) and is based on volumetric certainty criteria as well as on commercialization potential. For a detailed discussion of reserves and resources, refer to PRMS (2007, 2018).

4.3.3.3 Subsurface uncertainty: conclusions

Hopefully, we have shown that subsurface uncertainty remains significant up until the time when dynamic data from production history is available to update reservoir performance predictions and reduce that uncertainty. We have found that subsurface uncertainty is often underestimated or even "forgotten" after the FID has been made. If reservoir "surprises" arise that impact production, they are remedied as much as possible during operations. It is important to remember that remaining uncertainties at the FID stage are—or should be—close to "irreducible" given the available data and knowledge at that time.

Not all projects are equal. When a project incorporates high quality subsurface characteristics (reservoir and fluids) from the beginning, it can turn into a high-return project if the development is performed efficiently. However, if a project starts out with poor subsurface conditions (of the reservoir and/or fluids), it is unlikely it will exhibit stellar economics.

Key subsurface drivers that impact uncertainty around recoverable volumes are reservoir connectivity, complexity, size and quality. One source of uncertainty is the amount of heterogeneity leading to compartmentalization within the HC portion. This heterogeneity may be due to stratigraphic complexities or to structural reasons. In particular, fault transmissivity prediction is notoriously difficult in the absence of dynamic data. Several rules of thumb and/or workflows exist to predict fault transmissivity, including fault plane analyses and fault juxtaposition (see Chapter 3), but fault transmissibility is usually finalized and adjusted during reservoir simulations. It is important to keep in mind that understanding fault connections and whether or not they act as potential barriers or baffles is a difficult, non-unique iterative process at best that needs to incorporate dynamic data. Furthermore, certain faults that did act as barriers at initial reservoir conditions have been known to break down due to pressure depletion, such that they come to act more like baffles as production advances. Depending on the in situ stresses, the amount of baffling due to faults may need to be revisited as production occurs. For these reasons, static models usually incorporate a number of "engineering" faults (a.k.a. no-offset faults) initially assigned a transmissibility of 1 but which can be modified to allow for flexibility during simulation.

Another source of connectivity uncertainty is the amount of flow that might occur from aquifer to reservoir as the latter's pressure depletes through production. As is shown in Chapter 7, this is a function of aquifer geometry and strength. The information that can be inferred from a static model is the aquifer's volume and geometry, as well as an anticipated directionality of the aquifer influx.

REFERENCES

Allen P.A. and Allen, J.R. (2005). *Basin Analysis*. 2nd edn. Malden, MA: Blackwell.

Avseth, P. and Lehocki, I. (2016). Combining burial history and rock-physics modeling to constrain AVO analysis during exploration. *The Leading Edge* 33 (6): 528–534.http://dx.doi.org/10.1190/tle35060528.1.

Avseth, P. and Skjei, N. (2011). Rock physics modeling of static and dynamic reservoir properties—a heuristic approach for cemented sandstone reservoirs. *The Leading Edge* 30 (1): 90–96. http://dx.doi.org/10.1190/1.3535437.

Baur, F., Scheirer, A.H. and Peters, K.E. (2018). Past, present, and future of basin and petroleum system modeling. *AAPG Bulletin* 102 (4): 549–561.

Bogicevic, G., Dulic, I. and Sovilj, J. (2020). "3D Petroleum System Model of Southeastern Part of Pannonian Basin." *Search and Discovery* Article #11306. Available from www.searchanddiscovery.com/documents/2020/11306bogicevic/ndx_bogicevic.pdf .

Brown, A. (1997). Porosity variation in carbonates as a function of depth: Mississippian Madison Group, Williston basin. In J.A. Kupecz, J. Gluyas and S. Bloch (eds), *Reservoir Quality Prediction in Sandstones and Carbonates*, 29–46. AAPG Memoir 69. Tulsa, OK: AAPG.

Byrnes, A., Zhang, S., Canter, L. and Sonnenfeld, M. (2018). Application of integrated core and multiscale 3-D image rock physics to characterize porosity, permeability, capillary pressure, and two- and three-phase relative permeability in the Codell Sandstone, Denver Basin, Colorado. Unconventional Resources Technology Conference (URTeC), Houston, Texas, July 23–25, 3111–3130. https://doi.org/10.15530/urtec-2018-2901840.

Doyen, P. (2007). *Seismic reservoir characterization: An earth modelling perspective*. Houten: EAGE Publications.

Ehrlich, R., Kennedy, S.K., Crabtree, S.J. and Cannon, R.L. (1984). Petrographic image analysis, I: Analysis of reservoir pore complexes. *Journal of Sedimentary Research* 54 (4): 1365–1378. https://doi.org/10.1306/212F85DF-2B24-11D7-8648000102C1865D.

Evans, J., Cade, C. and Bryant, S. (1997). A geological approach to permeability prediction in clastic reservoirs. In J.A. Kupecz, J. Gluyas and S. Bloch (eds), Reservoir *Q*uality *P*rediction in *S*andstones and *C*arbonates, 91–101. AAPG Memoir 69. Tulsa, OK: AAPG.

Faizov, R., Maksimova, E. and Kolesnikov, D. (2019). Creating 3D geomodel: how to increase the quality through training and education? SPE Annual Technical Conference and Exhibition, Calgary, SPE-195821-MS.

Flemings, P. (2021). *A Concise Guide to Geopressure: Origin, Prediction, and Applications*. Cambridge: Cambridge University Press.

Giles, M.R. (1997). *Diagenesis: A Quantitative Perspective: Implications for Basin Modelling and Rock Property Prediction*. Dordrecht: Kluwer.

Gluyas, J. and Cade, C.A. (1997). Prediction of porosity in compacted sands. In J.A. Kupecz, J. Gluyas and S. Bloch (eds), Reservoir Quality Prediction in Sandstones and Carbonates, 19–27. AAPG Memoir 69. Tulsa, OK: AAPG.

Gluyas, J. and Swarbrick, R. (2004). *Petroleum Geoscience*. Malden, MA: Blackwell.

Isaak, E. and Srivastava, M. 1990. *An Introduction to Applied Geostatistics*. Oxford: Oxford University Press.

Kacewicz, M. and Xu, W. (2006). High-resolution prediction of rock properties and hydrocarbon chargethrough an integrated basin modeling/seismic inversion approach (abs.). AAPG International Conference and Exhibition, Perth, Western Australia, November 5–8, 2006. www.searchanddiscovery.com/pdfz/abstracts/pdf/2006/intl_perth/abstracts/ndx_kacewicz.pdf.html.

Magara, K. (ed.) (1978). *Compaction and Fluid Migration. Practical Petroleum Geology* 9: 1–319.

Major, R.P., and Holtz, M.H. (1997). Predicting reservoir quality at the Development scale: Methods for quantifying remaining hydrocarbon resource in diagenetically complex carbonate reservoirs.

In J.A. Kupecz, J. Gluyas and S. Bloch (eds), Reservoir *Q*uality *P*rediction in *S*andstones and *C*arbonates, 231–48. AAPG Memoir 69. Tulsa, OK: AAPG.

Mountjoy, E.W., and Marquez, X.M. (1997). Predicting reservoir properties in dolomites: Upper Devonian Leduc buildups, deep Alberta basin. In J.A. Kupecz, J. Gluyas and S. Bloch (eds), Reservoir *Q*uality *P*rediction in *S*andstones and *C*arbonates, 267–306. AAPG Memoir 69. Tulsa, OK: AAPG.

Mur, A. and Vernik, L. (2021). Rock physics modeling of silicious and carbonate facies: Conceptual AVO models applied to reservoirs around the globe. Proceedings of the 14th SEGJ International Symposium, Online, October 18–21, 238–242. https://doi.org/10.1190/segj2021-063.1.

Pittman, E.D. and Larese, R.E. (1991). Compaction of lithic sands: Experimental results and application. *AAPG Bulletin* 75 (8): 1279–1299.

Primmer, T.J., Cade, C.A., Evans, J., Gluyas, J.G., Hopkins, M.S., Oxtoby, N.H., Smalley, P.C., Warren, E.A. and Worden, R.H. (1997). Global patterns in sandstone diagenesis: Their application to reservoir quality prediction for petroleum exploration. In J.A. Kupecz, J. Gluyas and S. Bloch (eds), *Reservoir Quality Prediction in Sandstones and Carbonates*, 61–77. AAPG Memoir 69. Tulsa, OK: AAPG.

PRMS (2007). Petroleum Resources Management System, sponsored by Society of Petroleum Engineers (SPE), American Association of Petroleum Geologists (AAPG), World Petroleum Council (WPC) and Society of Petroleum Evaluation Engineers (SPEE).

PRMS (2018). Petroleum Resources Management System update, sponsored by World Petroleum Council (WPC), American Association of Petroleum Geologists (AAPG), Society of Petroleum Evaluation Engineers (SPEE), Society of Exploration Geophysicists (SEG), European Association of Geoscientists and Engineers (EAEG) and Society of Petrophysicists and Well Log Analysts (SPWLA).

Ramm, M., Forsberg, A.W. and Jahren, J.S. (1997). Porosity-depth trends in deeply buried Upper Jurassic reservoirs in the Norwegian Central Graben: An example of porosity preservation beneath the normal economic basement by grain-coating microquartz. In J.A. Kupecz, J. Gluyas and S. Bloch (eds), *Reservoir Quality Prediction in Sandstones and Carbonates*, 177–199. AAPG Memoir 69. Tulsa, OK: AAPG.

Saad, A., Khalaf, B. and Hassan, A. (2019). How to understand reservoir heterogeneity by using stochastic geomodels: Case study from Sequoi field, offshore Nile delta, Egypt. Paper presented at the Offshore Mediterranean Conference and Exhibition, Italy.

Schoener, R., Lueders, V., Ondrak, R., Gaupp, R. and Moeller, P. (2008). Fluid–rock interactions. In R. Littke, U. Bayer, D. Gajewski and S. Nelskamp (eds), *Dynamics of Complex Intracontinental Basins: The Central European Basin System*, 125–153. Berlin: Springer Verlag.

Shalaby, M.R., Abdullah, W.H. and Shady, A.N.A. (2008). Burial history, basin modeling and petroleum source potential in the Western Desert of Egypt. *Bulletin of the Geological Society of Malaysia* 54: 103–113.

Tissot, B.P. and Welte, D.H. 1984. *Petroleum Formation and Occurrence*. 2nd edn. Berlin: Springer Verlag.

Tobin, R.C. 1997. Porosity prediction in frontier basins: A systematic approach to estimating subsurface reservoir quality from outcrop samples. In J.A. Kupecz, J. Gluyas and S. Bloch (eds), *Reservoir Quality Prediction in Sandstones and Carbonates*, 1–18. AAPG Memoir 69. Tulsa, OK: AAPG.

Chapter 5

Aquifer analytical modeling

INTRODUCTION

This chapter discusses the most common theoretical solutions for aquifer influx into a reservoir caused by reservoir depletion. Readers are also provided with references to published reservoir engineering textbooks for derivations and/or for other interesting solutions to the aquifer influx problem.

Aquifer performance is controlled by both size (volume) and strength (transmissibility, total compressibility and baffling distributed over the aquifer geometry). Aquifer size is often reported as the ratio of aquifer volume (AQ) over reservoir hydrocarbon (HC) volume (AQ:HC). While no industry convention exists for describing aquifer strength, an aquifer strength index can be defined as hydrocarbon pore volume (HCPV) aquifer water encroachment per day per psi pressure drawdown per surface area at the free water level. This will be discussed in another chapter.

5.1 MASS AND ENERGY BALANCE

The general mass and energy balance in a petroleum reservoir can be depicted as in Figure 5.1.

As an equation (see Ahmed 2006), a petroleum reservoir's general mass and energy balance can be written as:

$$N_p B_o + \left(G_p - N_p R_s\right) B_g + W_p B_w = N\left(B_o - B_{oi}\right) + NB_{oi}\left(1+m\right)\left(\frac{c_w S_{wc} + c_f}{1 - S_{wc}}\right)\Delta P$$
$$+ N\left(R_{si} - R_s\right) B_g + mNB_{oi}\left(B_g / B_{gi} - 1\right) + W_e B_w \quad (5.1)$$

where
  B_o, B_g, B_w are formation volume factor for oil, gas and water respectively,
  N_p, G_p, W_p are cumulative oil, gas and water production respectively,
  N is the original in-place oil volume,
  R_s is solution gas–oil ratio,
  m is the ratio of initial gas-cap volume to initial oil-zone volume at reservoir conditions,
  c_w and c_f are water and formation compressibility,
  S_{wc} is connate water saturation,
  ΔP is depletion (pressure drop),

DOI: 10.1201/9781003272809-5

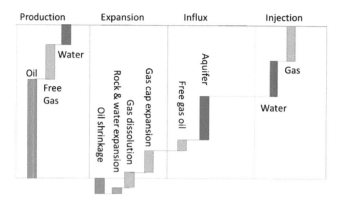

Figure 5.1 General mass and energy balance in a petroleum reservoir.

W_e is water influx from the aquifer,
 $_i$ designates initial conditions.

Note that the injection terms, $W_{inj}B_w + G_{inj}B_{g\,inj}$, are excluded from the right-hand side of the above equation.

Most reservoirs underlain by an aquifer require at least 15–25% pressure depletion before the aquifer's effect is detectable. Material balance methods are most reliable in high-permeability reservoirs with consistent pressure decline and an abundance of high quality data points. Aquifer pressure support must be accounted for if present. It is difficult to determine average reservoir pressures for low-permeability reservoirs with few available data points and/or with a poor correlation of pressure data points. It should be noted that classical material balance is zero-dimensional, i.e. it assumes constant pressure and properties at all points in the reservoir tank. As a result, it may not be applicable to reservoirs with high vertical relief and compositional variation. Material balance models can use multiple tanks with transmissibility multipliers to represent partial tank communication.

5.2 VAN EVERDINGEN AND HURST

Van Everdingen and Hurst (1949) used the following dimensionless parameters in order to solve the pseudo-steady state flow of a cylindrical aquifer surrounding a reservoir:

$$W_e = 2\pi \phi h \bar{c} \, r_o^2 \Delta \rho W_{eD}(t_D) \tag{5.2}$$

where
  W_e is water encroachment,
  $\phi, h, \bar{c}$ are aquifer porosity, thickness and compressibility respectively,
  r_o is the outer aquifer radius,
  $\Delta \rho$ is pressure drop,
  W_{eD} is the dimensionless water influx as a function of dimensionless time, i.e. t_D.

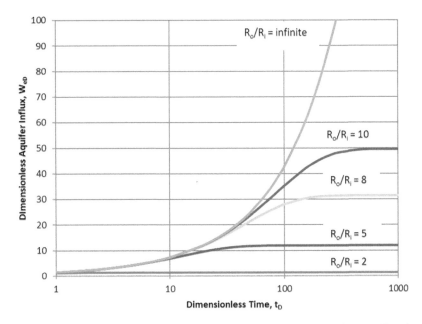

Figure 5.2 Dimensionless water influx for different aquifer sizes (modified after van Everdingen and Hurst 1949).

$$t_{D_j} = \frac{6.33kt_j}{\phi\mu_w c_{wr} r_o^2} \qquad (5.3)$$

The dimensionless type curves for water influx for different R_o/R_i (=r_e/r_R) are shown in Figure 5.2 (Ahmed 2006: 668). These curves indicate that permeability impacts the timing of the response, but the amount of water influx is dependent on porosity, compressibility and aquifer size only.

The van Everdingen and Hurst method (VEH) assumes constant flux at the reservoir boundary. For real problems with varying flux, superposition is used to combine multiple solutions in order to describe reservoir history. All aquifers initially act as if they were infinite in size but not in strength. Aquifer size can only be detected dynamically once the flow becomes boundary-dominated. If boundary-dominated flow has not been reached, history-matching can only identify the minimum aquifer size needed to match the pressure support observed in the history. In this situation, aquifer volumes that are both larger and further away should provide the same history-match. Long-term production forecasts can be pessimistic if the aquifer size is too small to continue to provide pressure support.

5.3 HAVLENA-ODEH METHOD

Havlena-Odeh (HO) (1963) converted Equation 5.1 into dimensionless groups, as can be seen below:

124   Integrated Aquifer Characterization and Modeling

$$F = N\left[E_o + mE_g + E_{fw}\right] + W_e \tag{5.4}$$

where

$$F = N_p\left[B_t + B_g\left(R_p - R_{si}\right)\right] + W_p B_w \tag{5.5}$$

$$B_t = B_o + B_g\left(R_{si} - R_s\right) \tag{5.6}$$

E_o, E_g and E_{fw} are the cumulative expansion of oil, gas, and of formation and water, respectively.

$$E_o = B_t - B_{ti} \tag{5.7}$$

$$E_g = B_{ti}\left[\frac{B_g}{B_{gi}} - 1\right] \tag{5.8}$$

$$E_{fw} = (1+m)B_{oi}\left[\frac{c_w S_{wi} + c_f}{1 - S_{wi}}\right]DP \tag{5.9}$$

By plotting

$$\frac{F}{E_o}\left(\text{or } \frac{F}{E_g} \text{ for a gas reservoir}\right) \text{ versus } \frac{W_e B_w}{E_o}\left(\text{or } \frac{W_e B_w}{E_g} \text{ for a gas reservoir}\right) \tag{5.10}$$

for various dimensionless radii—aquifer/oil radius (R_{eD})—a linear trend is expected for an active aquifer of a certain volume. Figure 5.3 illustrates a gas field example.

Water influx is calculated using VEH at various R_{eD} ratios of 5, 7 and infinite. This provides an initial input into Material Balance model ahead of (MBAL) water influx modeling for history-matching.

HO plot of $\frac{F}{E_g}$ versus $\frac{W_e B_w}{E_g}$ indicates an R_{eD} of 7 and a range of 2.4–3 trillion cubic feet (TCF) for volume of gas originally in place (OGIP), as shown by the y-intercept in Figure 5.3 (see Ahmed 2006: 796–798).

5.4 COLE PLOT FOR ESTIMATING AQUIFER CONTRIBUTION

The Cole plot is the basis for another method of distinguishing between depletion and water drive in a gas reservoir. In this method, the material balance equation is grouped as follows (Pletcher 2002):

$$\frac{G_p B_g}{B_g - B_{gi}} = G + \frac{W_e - W_p B_w}{B_g - B_{gi}} \tag{5.11}$$

As shown in Figure 5.4, different trend results are obtained depending on aquifer drive strength.

Aquifer analytical modeling 125

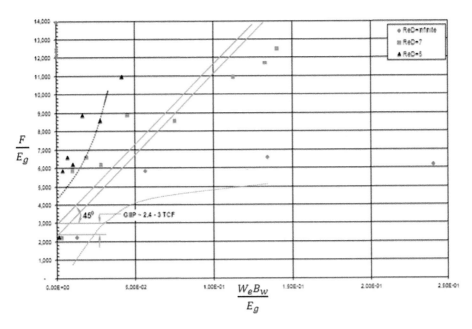

Figure 5.3 Havlena-Odeh method.

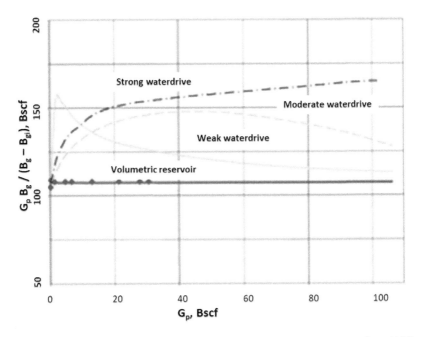

Figure 5.4 Cole plot curve shapes as a function of aquifer strength (modified after Cole 1969).

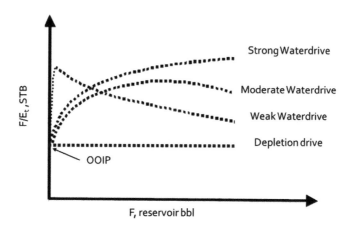

Figure 5.5 Campbell plot (modified after Campbell and Campbell 1978).

5.5 CAMPBELL PLOT FOR ESTIMATING AQUIFER CONTRIBUTION IN AN OIL RESERVOIR

Similar to the Cole plot, the Campbell plot method is used to distinguish between aquifer drive and depletion (Campbell and Campbell 1978). Figure 5.5 is an example of a Campbell plot.

$$\frac{F}{E_t} = N + \frac{W_e}{E_t} \tag{5.12}$$

where
  E_t = cumulative total expansion:

$$E_t = E_o + mE_g + E_{fw} \tag{5.13}$$

5.6 AQUIFER EFFECTIVENESS IN LOWER QUALITY GAS RESERVOIRS

In lower quality gas reservoirs, water permeabilities can be two to three orders of magnitude lower than air permeabilities in the area where water influx traps gas. The reduced water permeabilities can cause gas reservoirs with low absolute permeability, i.e. < 100 millidarcys (md), to behave volumetrically during a relatively short production period. Figure 5.6 illustrates the impact of this phenomenon. Notice that ultimate recoveries are equal for all cases.

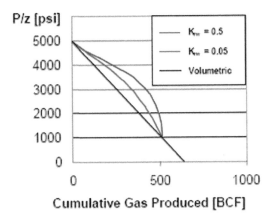

Figure 5.6 Material balance sensitivity to end-point relative permeability, i.e. Krw at Sw=1-Sgr (modified after Hower and Jones, 1991).

5.7 P/Z (PRESSURE/COMPRESSIBILITY) PLOT

A P/Z plot is generally used to estimate OGIP and cumulative producible gas (G) at a certain abandonment pressure. However, for a reservoir that has not been affected by an aquifer, there is more uncertainty with earlier P/Z data. An early linear trend in this plot could start curving once the aquifer becomes active, as shown in Figure 5.7.

5.8 SOLUTION PLOT FOR GAS RESERVOIRS

The material balance equations can be rearranged to arrive at Equation 5.14:

$$\frac{1}{P_i - P}\left(\frac{P_i/z_i}{P/z} - 1\right) = \frac{1}{G}\left[\frac{G_p}{(P_i - P)} \cdot \frac{P_i/zi}{P/z}\right] - \left[\frac{198.4(W_e - W_p)B_w P_i}{(P_i - P)Gz_i T} + \frac{S_{wi}C_w + C_f}{(1 - S_{wi})}\right] \quad (5.14)$$

Like the Cole plot, the solution plot method (Poston et al. 1994) theoretically defines the correct initial gas volume from early time data.

OGIP is derived from the reciprocal of the straight line drawn through or regressed to the data points. As with the Cole plot approach, scatter is likely to occur for early time data, resulting in a range of lines that could represent a good fit.

The solution plot method (see Figure 5.8) is more effective at estimating OGIP when permeabilities in the reservoir are relatively high.

128 Integrated Aquifer Characterization and Modeling

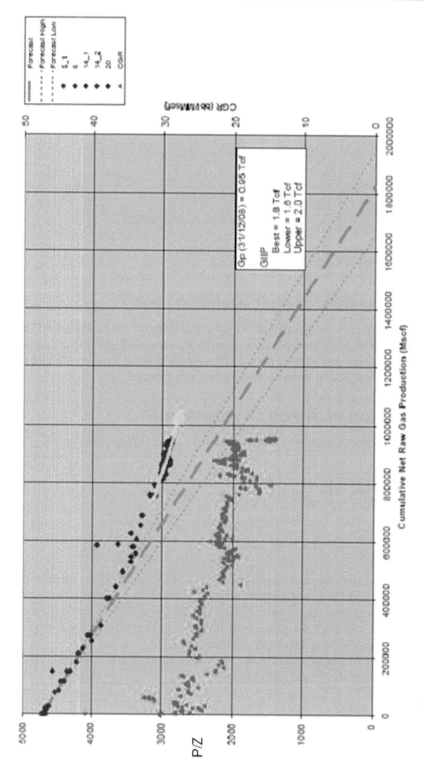

Figure 5.7 P/Z diagram for a gas reservoir with an active aquifer.

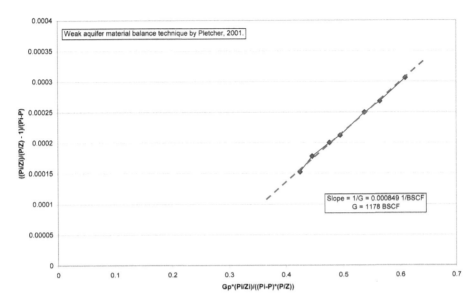

Figure 5.8 Solution plot.

5.9 CARTER-TRACY AQUIFER IMPLEMENTATION

The Carter-Tracy analytical method (Carter and Tracy 1960) offers the advantage of calculating the aquifer solution without superposition. Instead, this method requires an input of "influence functions" as a function of the ratio of the outer aquifer radius divided by the radius of the reservoir over aquifer interface (R_o/R_i). This technique's simplifying premise is to assume constant flux. As implied by this radius parameter, the analysis is one-dimensional in the r-direction.

The influence functions can be calculated based on a regression proposed by Fanchi (1985) as follows:

$$t_D = \frac{0.00634\ kt}{\varnothing \mu c R_o^2} \tag{5.15}$$

$$P_{tD} = a_0 + a_1 t_D + a_2 \ln(t_D) + a_3 \ln(t_D)^2 \tag{5.16}$$

where
  t_D is dimensionless time,
  P_{tD} is dimensionless pressure, and
  a_0, a_1, a_2 and a_3 are regression coefficients.

The influence functions in the regression above are accurate except for early times when they need to be corrected. After the P_{tD} values are calculated with the above formula, they should be compared with the values from the infinite case. If the value of the resulting calculated

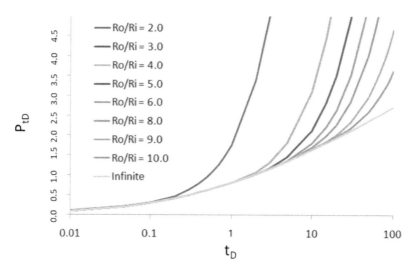

Figure 5.9 Carter-Tracy influence functions.

P_{tD} is greater than the infinite case value P_{tD}, then it should be replaced with the latter. The corrected curves are illustrated in Figure 5.9 (where $P_D = P_{tD}$), where one can see that all P_{tD} values for different R_o/R_i ratios have a minimum located at the curve for infinite R_o/R_i.

5.10 FETKOVICH AQUIFER

The Fetkovich aquifer method (1971) is another option offered by most numerical reservoir simulators. Dake (1978: 319–332) provides an extensive evaluation and summary of this method.

The Fetkovich method is appropriate for simulating small, quick-acting, highly permeable aquifers. It applies a pseudo steady-state productivity index and material balance between aquifer pressure and cumulative influx. Pressure response is assumed to be felt uniformly throughout the entire aquifer. Transient effects such as those at field start-up are not captured.

Dake's comparisons of the Fetkovich method to VEH show that the former is very accurate for $R_o/R_i < 5.0$ and somewhat less accurate for $R_o/R_i = 10.0$. Marques and Trevisan (2007) show that for $R_o/R_i > 12.0$, the Fetkovich method no longer yields acceptable results.

5.11 OTHER AQUIFER MODELS

Schilthuis (1936) assumes aquifer flux is proportional to ΔP.

Hurst (1943) assumes aquifer flux is proportional to depletion over time.

The small pot (Coats 1970) aquifer assumes uniform pressure depletion.

Vogt-Wang (Vogt and Wang 1990) is similar to VEH but uses a linear pressure decline at the reservoir–aquifer boundary for each timestep.

REFERENCES

Ahmed, T. (2006). *Reservoir Engineering Handbook*. 3rd edn. Cambridge, MA: Elsevier.

Campbell, R.A. and Campbell, J.M. (1978). *Mineral Property Economics. Volume 3.Petroleum Property Evaluation*. Norman, OK: Campbell Petroleum Series. www.osti.gov/biblio/5958322-mineral-property-economics-volume-petroleum-property-evaluation.

Carter, R.D. and Tracy, G.W. (1960). An improved method for calculating water influx. *Petroleum Transactions of the AIME* 219 (1): 415–417. SPE-1626-G. Available from https://doi.org/10.2118/1626-G.

Coats, K.H. (1970). Mathematical methods for reservoir simulation. Presented by the College of Engineering, University of Texas at Austin, June 8–12.

Cole, F.W. (1969). *Reservoir Engineering Manual*. Houston, TX: Gulf Publishing Company.

Dake, L.P. (1978). *Fundamentals of Reservoir Engineering*. New York: Elsevier Science Publishers.

Fanchi, J.R. (1985). Analytical representation of the van Everdingen-Hurst aquifer influence functions for reservoir simulation. *Society of Petroleum Engineers Journal* 25 (3): 405–406. SPE-12565-PA. https://doi.org/10.2118/12565-PA.

Fetkovich, M.J. (1971). A simplified approach to water influx calculations-finite aquifer systems. *Journal of Petroleum Technology* 23: 814–828. SPE-2603-PA. https://doi.org/10.2118/2603-PA.

Havlena, D. and Odeh, A.S. (1963). The material balance as an equation of a straight line. *Journal of Petroleum Technology* 15 (8): 896–900. SPE-559-PA. Available from https://doi.org/10.2118/559-PA.

Hower, T.L. and Jones, R.E. (1991). Predicting recovery of gas reservoirs under waterdrive conditions. Paper presented at the SPE Annual Technical Conference and Exhibition, October 6–9, SPE-22937-MS. https://doi.org/10.2118/22937-MS.

Hurst, W. (1943). Water influx into a reservoir and its application to the equation of volumetric balance. *Transactions of the AIME* 151 (1): 57–72. SPE-943057-G. Available from https://doi.org/10.2118/943057-G.

Marques, J. B. and Trevisan, O.V. (2007). Classic models of calculation of influx: a comparative study. Paper presented at the Latin American & Caribbean Petroleum Engineering Conference, Buenos Aires, April. SPE-107265-MS. https://doi.org/10.2118/107265-MS.

Pletcher, J.L. (2002). Improvements to reservoir material balance methods. *SPE Reservoir Evaluation & Engineering* 5 (1): 49–59. SPE-75354-PA. https://doi.org/10.2118/75354-PA.

Poston, S.W., Chen, H.Y. and Akhtar, M.J. (1994). Differentiating formation compressibility and water-influx effects in overpressured gas reservoirs. *SPE Reservoir Engineering* 9 (3): 183–187. https://doi.org/10.2118/25478-PA.

Schilthuis, R.J. (1936). Active oil and reservoir energy. *Transactions of the AIME* 118 (1): 33–52. SPE-936033-G. Available from https://doi.org/10.2118/936033-G.

Van Everdingen, A.F. and Hurst, W. (1949). The application of the Laplace transformation to flow problems in reservoirs. *Journal of Petroleum Technology* 1 (12): 305–324. SPE-949305-G. Available from https://doi.org/10.2118/949305-G.

Vogt, J.P. and Wang, B.A. (1990). More accurate water influx formula with applications. *Journal of Canadian Petroleum Technology* 29 (4): paper no. PETSOC-90-04-04. https://doi.org/10.2118/90-04-04.

Chapter 6

Numerical aquifer modeling

SUMMARY

As has been discussed previously, aquifer performance is controlled by both size (volume) and strength (transmissibility, total compressibility and baffling distributed over aquifer geometry). In most cases, the previously discussed analytical aquifer models (Chapter 5) are too simple to adequately describe real aquifers. Chapter 6 will show, for better accuracy, an aquifer can be modeled as a series of numerical aquifer "tanks" whose connectivity can be estimated using the dynamically measured data in the field. Until an aquifer operates in boundary-dominated flow, production history-matching of the reservoir can only determine the minimum aquifer volume and its strength. Maximum aquifer volume is obtained from mapping. As will be discussed in Chapter 9, when aquifers are used for gas storage, detailed gridded models are required.

INTRODUCTION

There are two basic approaches to the numerical representation of aquifers in dynamic models. The first method is to define relatively few—but large—grid cells connected through non-neighbour connections. This representation is called the "numerical aquifer."

A second gridding method is the "conventional" approach of extending the grid beyond the reservoir and defining the properties of the aquifer in the same way that the reservoir grid cells are defined. Aquifer cells near the reservoir boundary are about the same size as reservoir cells, but usually become significantly larger in the direction away from the reservoir. These two methods are discussed in detail in this chapter.

6.1 REQUIRED INFORMATION

The following data is needed prior to doing any numerical aquifer modeling:

- map of aquifer extent
- data on any communication barriers/baffles between aquifer extent and completed intervals in production wells
- aquifer properties (permeability, porosity, compressibility) and their variations areally and vertically
- water–oil contacts (or water–gas contacts for gas reservoirs)
- end-point relative permeability, particularly water mobility and displacement efficiency

- fluid saturations as functions of depth (including in the aquifer)
- gas dissolution in the aquifer brine, if any

6.2 SOURCES OF INFORMATION

This data can be obtained from several sources. A partial list is included below:

- high quality seismic imaging and interpreted products
- well penetrations with high quality logging and special core analysis (SCAL) testing
- static and dynamic pressure surveillance
- reservoir fluid sampling and production validation testing
- production history
- reservoir and aquifer property geomodeling and Touchstone modeling
- properties and performance of analogue reservoirs

6.3 NUMERICAL AQUIFER

The numerical aquifer option treats cells as conventional grid cells[1], only larger. In order to save computation time, relatively few cells are used to account for aquifer volumes. Communication among the cells is made through non-neighbour connections. Modifying transmissibilities among cells offers some flexibility for managing aquifers' variability. This allows for a coarse representation of geological features such as faults, permeability barriers, changing rock properties, etc. and constitutes a significant advantage of the numerical approach as compared to the Carter-Tracy method, which does not account for communication between aquifers.

As with conventional gridding, numerical accuracy can be improved by refining grid size. Figure 6.1 illustrates how grid cells should be coarsened gradually in the direction of the outer aquifer. The numerical error called "numerical dispersion" is proportional to the grid size. If grid cells are smaller in the area of interest and only gradually increase in size away from the main part of the reservoir, the numerical method does a better job of representing flow equations.

Gridding in the direction parallel to the boundary between the reservoir and the aquifer can be another significant source of numerical errors. Figure 6.2 depicts an example where many cells in the regular grid connect to a common numerical aquifer cell. This numerical cell contains a large volume internally that facilitates instantaneous communication with all of the connecting grid cells. Any faults, barriers or baffles in the grid near the aquifer boundary are effectively bypassed. The instantaneous connection among all cells—shown in pink in Figure 6.2—could significantly alter the aquifer's impact on reservoir performance.

Figures 6.3 and 6.4 illustrate a more physically realistic numerical aquifer that allows connections to more aquifer cells than is the case in Figure 6.2. Instead of having just one

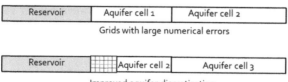

Figure 6.1 Gridding of an aquifer.

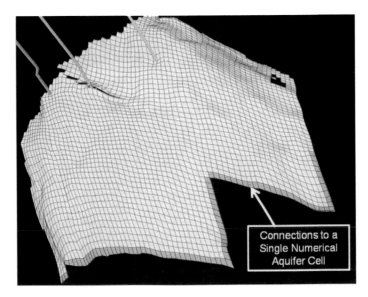

Figure 6.2 Single numerical aquifer.

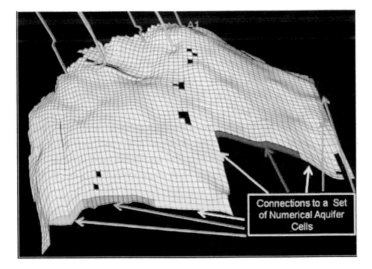

Figure 6.3 Multiple numerical aquifers.

numerical aquifer cell connecting to the reservoir, eight cells have been created along the reservoir boundary. In Figure 6.4, four additional aquifer cells are present in the direction away from the reservoir–aquifer interface.

There were 32 (i.e. 8×4) numerical aquifer cells in this geological stratum overall. Two additional geological zones were present under the one shown, so a total of 96 aquifer cells were active for all three zones. This type of numerical discretization can do a better job of representing the interaction between aquifer and reservoir.

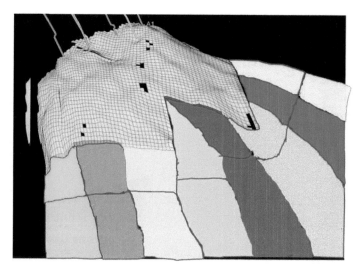

Figure 6.4 Multiple numerical aquifers (fully extended).

6.4 CONVENTIONAL GRIDDED AQUIFER

The conventional grid aquifer model uses conventional grid cells to represent the aquifer. The cell size is the same as reservoir cells near the reservoir boundary and gradually becomes larger away from the reservoir. Thus, detailed geological features such as faulting, properties' changes, and geometries can be accounted for. The main disadvantage of this model is that it requires many more grid cells, which increases the run time of the simulation. If knowledge of the aquifer is limited—as is often the case—then the extra effort involved in adding cells to the grid may well not be warranted.

The conventional aquifer approach is recommended for smaller aquifers (AQ:HC < 5), or if geological features in the aquifer area are known. Conventional cells can also model gas dissolved in the aquifer. Otherwise, the numerical aquifer option is more suited and generally represents aquifer behavior more efficiently than the conventional method.

6.5 AQUIFER ANALYTICAL MODELS IN RESERVOIR SIMULATORS

Aquifer influx into the reservoir at the interface between the two is generally modeled using the analytical functions (see Chapter 5). These functions calculate the aquifer volume that flows into the simulation grid as a function of (a) pressure conditions at the reservoir boundary, and (b) time since the reservoir started producing. Existing reservoir simulators have the following options for achieving this purpose.

6.5.1 Carter-Tracy aquifer guidelines

The Carter-Tracy (Carter and Tracy 1960) aquifer approach has been applied extensively in numerical simulation models and has proven to be an effective tool. There are, however, several limitations that users should be aware of and take into account when creating dynamic simulation models using this method. Our observations here are based on personal and colleagues' experience as well as on information from the petroleum literature and from numerical simulators manuals.

The Carter-Tracy method provides good accuracy in accounting for aquifer behavior in the radial direction away from the reservoir, as it was designed to do. While the numerical-analytical interface at the boundary has not been a problem, there are accuracy issues that need to be addressed.

One of the limitations of this method is that Carter-Tracy aquifers do not communicate with each other; they only communicate with the reservoir at the specified connections. If multiple Carter-Tracy aquifers are applied, they will act fully independently—which may not be the case in the field. This restriction points to another limitation of this approach, i.e. that it cannot account for geological information in the aquifer. For example, the effect of faulting or other potential barriers is not included in its results. Carter-Tracy is valid only for a simple, one-dimensional aquifer representation.

In a 2007 aquifer evaluation study, Marques and Trevisan indicated that when compared to a rigorous analytical solution, the accuracy of the Carter-Tracy approach degrades for aquifers with a $R_e/R_o < 5.0$. This conclusion is based on the analytical results presented in their paper but has not been tested in the numerical examples here.

The use of Carter-Tracy aquifers also entails practical limitations. Reservoir simulators should allow for aquifer volumes, influence tables and other aquifer features to be printed. They should also account for gas dissolution in the aquifer. Depending on the level of pressure depletion, gas dissolution could significantly impact the influx rate.

Lastly, when using analytical models, users often apply polygons in the mapping section to generate connecting reservoir cells. This part of the process requires a great amount of caution lest incorrect cells be generated. Under some reservoir conditions, an erroneous boundary could enhance aquifer impact significantly.

6.5.2 The Fetkovich aquifer

Another analytical method for simulating aquifer behavior is the Fetkovich approach. It is generally recommended for small, quick-acting, highly permeable aquifers. It applies a pseudo steady-state productivity index and material balance between aquifer pressure and cumulative influx. Pressure response is assumed to be felt uniformly throughout the entire aquifer. Transient effects such as those at field start-up are not captured.

Dake (1978: 319–332) compared the Fetkovich method to the van Everdingen-Hurst (VEH) approach, concluding that the former is very accurate for $R_e/R_o < 5.0$. Accuracy degrades somewhat for $R_e/R_o = 10.0$. Marques and Trevisan (2007) showed that, for the set of problems investigated in their paper, the accuracy of the Fetkovich method is not acceptable where $R_e/R_o > 12.0$. In general, we do not recommend the Fetkovich aquifer model because its validity is restricted to a narrow set of conditions and because its application can easily lead to misinterpretation.

Table 6.1 provides a comparison between different models representing aquifer behavior in a reservoir simulator.

Table 6.1 Capabilities comparison of analytical and numerical aquifer models

Aquifer option	Comparison of aquifer models capabilities						
	Geology	Physics	Transparency	Grid dependency	Accuracy	Ease of use	Recommendation
Carter-Tracy	1	3	3	4	3	3	3
Fetkovich	1	2	2	4	2	4	1
Numerical	2	3	4	3	3	3	4
Conventional	3	4	3	3	3	2	3

Note: 1 = poor or non-existent; 2 = marginal; 3 = adequate; 4 = good

6.6 GUIDELINES FOR ANALYTICAL MODEL USE

The following guidelines help inform the use of the analytical models discussed in this chapter:

(a) *Carter-Tracy*
- is a valid and often effective approach to aquifer modeling
- is a less transparent approach that can be easily misapplied
- is only applicable to the connection between aquifer and reservoir boundary; the fact that aquifers are not technically connected to each other may be a limitation in some models
- is noted for its loss of accuracy for $R_o/R_i < 5$

(b) *Fetkovich*
- is not a transparent approach within certain parameters
- is only applicable to small aquifers

(c) *Numerical*
- is the most transparent of all the methods
- entails the application of normal gridding rules, affecting solution accuracy
- is the most flexible approach; can allow for geological features with coarse approximation
- uses conventional gridding when practical

6.7 USING MBAL TO VALIDATE AN AQUIFER MODEL

The material balance software toolkit MBAL is often used to validate effective properties in numerical or conventional aquifers. MBAL affords the application of superposition, the mathematical technique used by van Everdingen and Hurst in their analysis (1949). The following steps are recommended:

- extracting relevant pressures and cumulative production volumes from the numerical model
- developing the MBAL model by:
  ○ entering HC parameters, i.e. original oil in place (OOIP), original gas in place (OGIP), pressure, porosity, compressibility

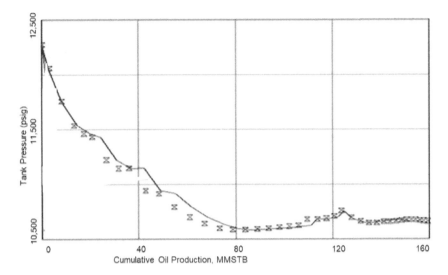

Figure 6.5 Regression match in MBAL using numerical simulator inputs.

- ○ entering aquifer parameters, i.e. radial aquifer inner and outer radius, time constant, aquifer constant (see Equations 5.2 and 5.3)
- comparing analytical and actual performance
- showing dimensionless performance

Once the results of pressure calculated as a function of time are imported into the MBAL model, a regression can be carried out in order to arrive at the right aquifer parameters. A good match with the analytical model in Figure 6.5 shows aquifer transients are well represented.

Aquifers are traditionally measured by their size relative to that of the HC volume. This procedure is a way to verify that the numerical aquifer correctly represents dynamic aquifer behavior.

6.8 KEY ISSUES IN AQUIFER MODEL BUILDING

- *MBAL aquifer size*: Aquifer volume is the sum of the water volume in the connected numerical aquifer "tanks" and the water volume below the free water level (FWL) in the HC reservoir tank(s). The HC volume is the HC present in the reservoir tanks communicating with the aquifer.
- *MBAL aquifer strength*: Aquifer encroachment is the sum of water movement from connected numerical aquifer "tanks" and water leg expansion of the HC reservoir tank(s). This encroachment volume is then normalized by hydrocarbon pore volume (HCPV), number of days in the reporting period and HC reservoir pressure drawdown.
- *Simulator aquifer size*: Aquifer volume is the sum of the water volume in the gridded and numerical aquifer cells in communication with the HC reservoir(s). HC volume

140 Integrated Aquifer Characterization and Modeling

is the sum of HC present in the corresponding reservoir intervals in communication with the gridded and numerical aquifer cells.
- *Simulator aquifer strength*: Aquifer encroachment is the sum total of water volume moving into gridded HC reservoir cells. This encroachment volume is then normalized by HCPV, number of days in the reporting period and HC reservoir pressure drawdown.
- A single large numerical aquifer cell can place large volumes of brine unrealistically close to a reservoir boundary. The better alternative is to use a series of numerical aquifer cells arranged to reflect mapped aquifer volumes.
- A single numerical aquifer can be connected along an extensive reservoir model boundary, providing instantaneous communication over very large distances. Instead, it is better to use multiple numerical aquifers and to check the connections to reservoir boundaries.
- *Software (a Schlumberger product)*: Petrel can generate erroneous connections between numerical aquifers and interior reservoir grid blocks. As a result, generating aquifer connections with Petrel should either be avoided entirely, or the results should be corrected by hand.
- History-matching often results in a minimum aquifer size. This may also show up as a drop in reservoir pressure at the end of production history, possibly resulting in pessimistic long-term forecasts. A better option is to use an aquifer size ranging from the low minimum history-matched size to the higher mapped aquifer size.
- Aquifer rock properties are too often assumed to be identical to reservoir rock properties in the absence of well penetrations. In some cases however, aquifer properties differ substantially from properties in the HC column and degrade significantly with depth. Therefore, only the best geoscience estimates of aquifer properties and their depth trends should be used.

6.9 ESTIMATION OF AQUIFER–RESERVOIR CONNECTIVITY

Aquifer connectivity to the reservoir is a key uncertainty in aquifer dynamic modeling. Once grid blocks are set up for an aquifer, it is important to verify how they are connected.

A long-term pressure buildup in a well can be used to derive both aquifer size and its connectivity. The impact of different ranges of connectivity is shown in Figure 6.6. The end-point in the pressure buildup curve represents the size of the aquifer, while the rate of pressure buildup represents the quality of its connectivity to the reservoir. In Figure 6.6, curve *a* represents a fully connected aquifer wherein the well bottomhole pressure stabilizes quickly upon shut-in. Curve *b* represents an aquifer with intermediate connectivity wherein the rate of pressure buildup is slower. Finally, curve *c* represents a large, poorly connected aquifer wherein pressure builds up much slower due to many tortuous paths and baffles. In fact, pressure has yet to stabilize for curve *c*, indicating a much larger aquifer.

Figure 6.7 depicts bottomhole pressure (BHP) buildup and its derivative for a GoM field during a long shutdown period of more than 500 days. Notice that the pressure continued to rise even after the long shutdown, indicating the presence of a tortuous and baffled path between aquifer and reservoir. As mentioned before, final pressure generally indicates aquifer size, and the right choice of AQ:HC ratio provides a good match for the final pressure. The pressure derivative—represented by $\Delta P/\Delta t$ in Figure 6.7—should be matched by adjusting the transmissibilities between aquifer grid blocks and reservoir. These issues will be discussed further in our field case studies in Chapter 8.

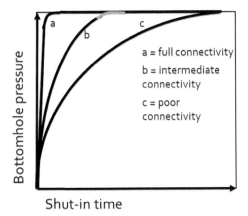

Figure 6.6 Pressure buildup behaviour for different aquifer types.

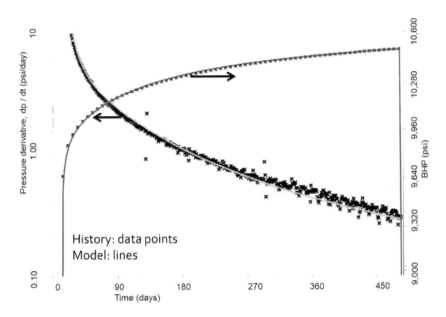

Figure 6.7 Bottomhole pressure buildup and its derivative.

6.10 CONCLUSIONS

- Aquifer performance is controlled by both size (volume) and strength (transmissibility, total compressibility and baffling distributed over aquifer geometry).

- In most cases, analytical aquifer models are too simple to adequately describe real aquifers. For better accuracy, an aquifer can be modeled as a series of numerical aquifer "tanks."

- Until an aquifer operates in boundary-dominated flow, history-matching can only determine the minimum aquifer volume and strength. Maximum aquifer volume is obtained from mapping.

NOTE

1 We are indebted to Dr. Martin Cohen for the concepts in this section and his initial study of aqifer modeling.

REFERENCES

Ahmed, T. (2006). *Reservoir Engineering Handbook*. 3rd edn. Cambridge, MA: Elsevier.

Carter, R.D. and Tracy, G.W. (1960). An improved method for calculating water influx. *Petroleum Transactions of the AIME* 219 (1): 415–417. SPE-1626-G. Available from https://doi.org/10.2118/1626-G.

Dake, L.P. (1978). *Fundamentals of Reservoir Engineering*. New York: Elsevier Science Publishers.

Fanchi, J.R. (1985). Analytical representation of the van Everdingen-Hurst aquifer influence functions for reservoir simulation. *Society of Petroleum Engineers Journal* 25 (3): 405–406. SPE-12565-PA. https://doi.org/10.2118/12565-PA.

Fetkovich, M.J. (1971). A simplified approach to water influx calculations-finite aquifer systems. *Journal of Petroleum Technology* 23 (7): 814–828. SPE-2603-PA. https://doi.org/10.2118/2603-PA.

Klins, M.A., Bouchard, A.J. and Cable, C.L. (1988). A polynomial approach to the van Everdingen-Hurst dimensionless variables for water encroachment. *Society of Petroleum Engineers Journal* 3 (1): 320–326. SPE-15433-PA. https://doi.org/10.2118/15433-PA.

Marques, J.B. and Trevisan, O.V. (2007). Classic models of calculation of influx: A comparative study. Paper presented at the Latin American & Caribbean Petroleum Engineering Conference, Buenos Aires, April. SPE-107265-MS. https://doi.org/10.2118/107265-MS.

Pletcher, J.L. (2000). Improvements to reservoir material-balance methods. SPE Annual Technical Conference, Dallas, Oct 1–4, SPE 75354.

Pletcher, J.L. (2002). Improvements to reservoir material-balance methods. *SPE Reservoir Evaluation & Engineering* 5 (1): 49–59.

Van Everdingen, A.F. and Hurst, W. (1949). The application of the Laplace transformation to flow problems in reservoirs. *Journal of Petroleum Technology* 1 (12): 305–324. Available from https://doi.org/10.2118/949305-G.

Chapter 7

Aquifer influx versus water injection in the Gulf of Mexico

INTRODUCTION

When developing new deep-water fields in locations such as the Gulf of Mexico (GoM), aquifer size and connectivity are two important variables that dictate the presence or absence of any future water injection project. Because of the substantial scale of the investment associated with developing fields in deep offshore environments, it is important to identify the key factors driving decisions on development optimization, preventing operators from incurring unnecessary expenses. The first such factor is the assessment of the aquifer's potential size and connectivity. The second factor is understanding the aquifer's strength. This second parameter has not been sufficiently assessed in the literature, so we propose a new aquifer influx index (AII) in order to address this gap.

Prior to any petroleum field start-up, it is important to plan for waterflooding, especially if the aquifer's strength is uncertain. We use data from several GoM fields, as well as their corresponding AII, so as to provide guidance on designing an effective conceptual water injection project prior to start-up.

7.1 AQUIFER DESCRIPTION WORKFLOW

Subsurface studies discussed in this chapter incorporate geophysical, geological and petrophysical activities that pertain to identifying and characterizing the static nature of aquifers and their connectivity to the reservoirs.

7.1.1 Geophysical input to aquifer size

Seismic is used to define the boundaries and the areal extent of aquifers, but it can also serve to assess the thickness of the aquifer and its potential internal complexity. Internal compartmentalization is often caused by faulting and/or by stratigraphic complexity within the reservoir interval. One of the tasks of the geophysicist is to define the limits of the container by creating ICE maps. The interpreter identifies potential edges to the trap (top, bottom, lateral), as well as any fluid contacts (GOC, OWC) detected by seismic and wells. After all edges are defined, the enclosed gross rock volume (GRV) for the given interval(s) associated with various segments can be calculated. The GRV computation does not entail an account of the properties of the rock contained within the volume. Instead, the surface of the area under consideration is multiplied by its average gross vertical thickness:

$$GRV = A.h \tag{7.1}$$

where A is the area of a rock body (in acres or ha), and h is the average vertical thickness of the rock body (in ft or m) in that area; h is sometimes also called true vertical thickness (TVT).

A segment is defined as a body of rock that forms a connected volume or flow unit. If multiple segments are present, then the overall GRV for the area in question is given by the sum of GRV values calculated for each segment:

$$GRV = \Sigma \ (A.h)_i \qquad (7.2)$$

Once the structural-stratigraphic interpretation is complete, the geophysicist may perform a quantitative seismic analysis by using a rock physics-based seismic elastic inversion to discriminate facies, as well as rock and fluid properties. The seismic inversion enables the assignment of elastic properties away from wells, which in turn allows for the classification of seismic facies and the development of a rock physics template (RPT) for rock property assignment to each facies. The structural and stratigraphic framework derived on a seismic grid provides an initial view of the potential complexity and connectivity of the geologic units in a given area.

Aquifers, like HC reservoirs, can exhibit seismic variation—in velocity, amplitude and frequency—due to changes in facies, porosity and/or pressure. An example of such variation is shown in Figure 7.1 (modified after Connolly 2010, personal communication), where the represented aquifer exhibits complex channel geometries within a deep-water turbidite system. Below the OWC, marked as a black line, the aquifer is complex, with various channels, distributaries and lobes.

Once a field is put into production, geophysicists interpret 4D seismic, which consists of 3D seismic surveys conducted periodically at various production intervals. When interpreted in conjunction with pressure history-matching from well data and reservoir simulation, 4D seismic is a key tool for the "monitoring and verification" of production processes such as water movement and/or changes in pressure. An integrated surveillance approach that incorporates all subsurface disciplines generates a valuable account of subsurface volumetric changes occurring during production that includes verification/reconciliation at the segment scale.

7.1.2 Geological and petrophysical input to aquifer effectiveness

Incorporated geological and petrophysical interpretations generate a static model of the subsurface that serves as the basis for a shared earth model (SEM). The SEM is subsequently augmented by dynamic information (fluid properties, pressure and fluid flow) to represent and test our understanding of the Earth's subsurface through time. This structural and stratigraphic information is mapped onto a reservoir simulation grid describing flow units together with expected seals and baffles. The geologist will attempt to characterize the transmissivity of faults and the amount of lateral continuity between bodies (or flow units) without much available dynamic data. Rock properties are assigned to all bodies present in the surveyed area, both reservoirs and non-reservoirs.

At this stage, special focus is given to the following properties: net-to-gross (NtG), fluid saturation, porosity, permeability, as well as any expected variation with depth, pressure and temperature. The geomodel attempts to capture lateral variations away from wells, as well as vertical heterogeneity at different scales. Transition zones (if present) due to capillary

Figure 7.1 3D perspective of a turbidite channel sand oil-aquifer system in deep water (based on Connolly 2010. SEG Spring Distinguished Lecture. SEG©2010, *reprinted by permission of the SEG whose permission is required for further use*).

effects are defined around HC contacts. With this information in hand, a preliminary measure of aquifer effectiveness can be approximated from the amount of water within the net rock porous volume (sometimes called storage). It is calculated as follows:

$$\text{Storage* } Sw \text{ Net rock porous volume* } Sw = \text{NRPV} = A.h.NtG.\phi.S_w \tag{7.3}$$

where ϕ is porosity (fraction) and S_w is water saturation.

If no HC is present and the saturation is 100%, then the last term, S_w, drops out of the equation. Aquifer effectiveness is expected to increase with net rock porous volume and with aquifer pressure. Conversely, aquifer effectiveness is expected to decrease with increasing reservoir complexity due to faulting and/or stratigraphic variations.

146 Integrated Aquifer Characterization and Modeling

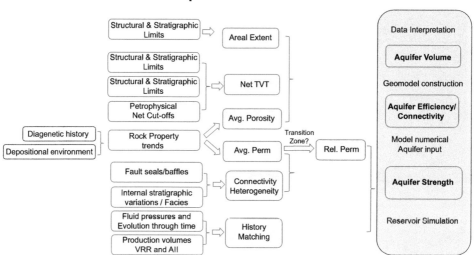

Figure 7.2 Recommended reservoir and aquifer characterization workflow, with three main components centered around aquifer volume, connectivity and strength.

Figure 7.2 illustrates our recommended workflow for characterizing HC reservoirs and their aquifers. The workflow is centered around three main components—aquifer volume, connectivity and strength—all of which are expected to impact reservoir performance. (Please see nomenclature for parameters in Figure 7.2 below.)

7.1.3 Static uncertainty of the aquifer

It is important to keep in mind that despite best efforts in characterizing aquifers, a significant amount of uncertainty is still associated with their calculated volume and connectivity. For example, a transition zone (e.g. between oil and water) requires both a deep understanding of local capillary effects and the modelling of its relative permeability using multi-phase flow. A tar mat will create additional uncertainty regarding aquifer connectivity.

$$OWIP = 7758.A.h.NtG.\phi / B_w \tag{7.4}$$

where OWIP is original water in place.

Figure 7.3 is an example of size uncertainty estimation for the aquifer shown in Figure 7.1 that assumes there is no transition zone between oil and water, and that no gas is dissolved in the aquifer. The inputs use an oil field approach (equation 7.4), where the water formation volume factor (B_w)—defined as the change in water volume in the formation as water moves from reservoir to surface conditions—is assumed to be close to 1. Assuming that most input distributions are Gaussian-like, the estimation yields a lognormal-like output. A total of 10,000 trials were used to illustrate the uncertainty propagation. The top right of the figure shows the output frequency distribution, while the bottom right shows the output exceedance

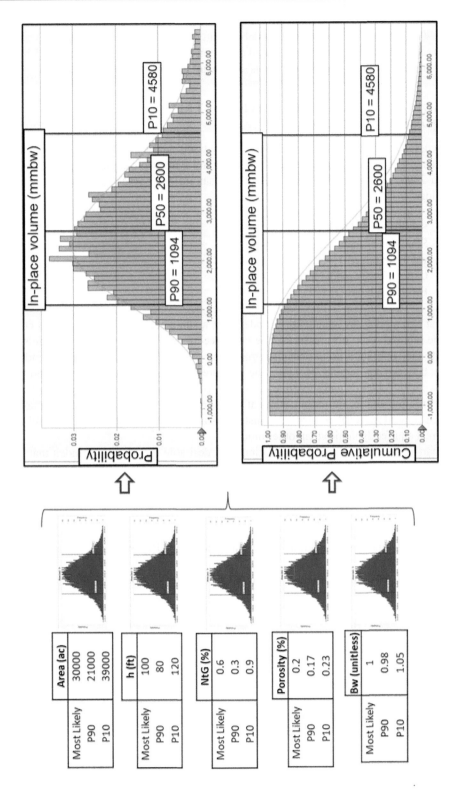

Figure 7.3 Calculated aquifer volumetric uncertainty, assuming input parameters are Gaussian-like.

curve. The expected volume (50% probability) is 2.6 billion (bn) barrels of water (bnbw), with a 90% chance that the water volume be greater than 1.09 bnbw and a 10% chance that it be greater than 4.58 bnbw. The standard deviation around a mean volume of 2.75 bnbw is approximately 1.4 bnbw, meaning there is significant volumetric uncertainty.

7.2 AQUIFER CHARACTERIZATION: STRENGTH QUANTIFICATION

As discussed in Chapter 5, analytical models have traditionally been used to estimate aquifer influx and have been discussed extensively in many a reservoir engineering textbook (e.g. Ahmed 2006). Among the popular models for saturated oil reservoirs, we can name the following:

- Schilthuis model (1936)
- Small- or pot-aquifer model (1970)
- van Everdingen-Hurst (VEH) (1949)
- Carter-Tracy model (1960)
- Fetkovich model (1971)

The first two models are steady-state models that assume aquifer pressure remains constant or is the same as reservoir pressures. The last three models are unsteady-state models. They simulate the complex pressure changes that occur gradually both in the aquifer and in the interval between aquifer and reservoir. The pressure difference between reservoir and aquifer increases with depletion and subsequently slows down when the aquifer and reservoir equilibrate. Thus, the aquifer influx rate starts at zero, grows steadily until it reaches a maximum and then slows down back to zero. The unsteady-state models are far more successful at capturing real aquifer dynamics than other models.

As part of aquifer strength quantification, it is desirable to arrive at an index that can be used for assessing aquifer connectivity. We have already discussed two important parameters, i.e. size and effectiveness.

One option is to use the *voidage-replacement ratio* (VRR) as a measure of aquifer effectiveness:

$$\text{VRR} = \text{cumulative water influx} / \text{cumulative produced HC and water} \quad (7.5)$$

where all parameters are calculated at reservoir conditions and the volume of injected water is excluded.

$$VRR = \frac{Q_{waq} \cdot B_w}{Q_{wp} \cdot B_w + Q_{op} \cdot B_o} \quad (7.6)$$

where Q_{waq} is cumulative aquifer influx, and Q_{wp}, Q_{op} are cumulative produced water and oil, respectively; B_w, B_o are water and oil formation volume factor, respectively.

A better option is to resort to a definition similar to the one used to measure the well productivity index, i.e. the *specific aquifer influx index* (SAII):

$$\textit{SAII} = \text{cumulative water influx} / (\text{time} \times \text{depletion} \times \text{effective surface area}) \quad (7.7)$$

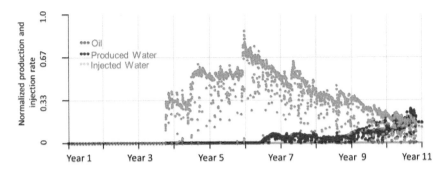

Figure 7.4 Production and injection rate for a strong aquifer (reservoir A3).

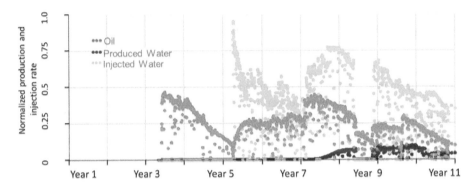

Figure 7.5 Production and injection rate for a poor aquifer (reservoir A4).

While effective surface area is easy to estimate for edge water drive, the same is not true in other cases. Furthermore, depletion is impacted by the amount of injected water for reservoirs in the water injection phase. For ease of application, we propose the following definition for **aquifer influx index** (AII):

$$AII = \frac{Q_{waq} \cdot B_w}{t \cdot \Delta p} \tag{7.8}$$

where
t is producing time (days),
Δp is depletion (psi or kPa),
AII is aquifer influx index (reservoir bbl/day/psi or reservoir m³/day/kPa).

Typical reservoir performance behavior under a strong aquifer (Reservoir A3) and a poor aquifer (Reservoir A4) are shown in Figures 7.4 and 7.5 respectively. The strong reservoir performance in reservoir A3 prior to water injection and the low injected water are indications

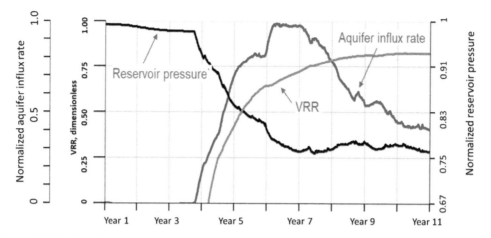

Figure 7.6 Aquifer influx, VRR and reservoir pressure for reservoir A3.

of a strong aquifer (AQ:HC ~ 250). However, in reservoir A4, the small aquifer (AQ:HC ~ 0.5) had to be augmented by a strong water injection in order to sustain production.

As part of history-matching the performance of several GoM oil fields, key aquifer parameters such as aquifer influx volume were calculated. Figure 7.6 displays the history-matched reservoir pressure, VRR and aquifer influx rate for reservoir A3. During the first three years, there was a shallow decline in reservoir pressure due to other wells' interference with production. Once the depletion rate increased, then aquifer influx also increased. As a result, VRR approached unity, indicating that aquifer influx closely matched the depletion rate in this reservoir.

Similar data were obtained via history-matching for 11 reservoirs within 3 fields in the GoM (Fassihi and Blangy, 2022). For brevity, those results are not shown here. Table 7.1 displays the estimated **VRR** and **AII** for these 11 reservoirs (Fassihi abd Blangy, 2022).

The aforementioned **AII** definition was used to arrive at the categorization in Figure 7.7. This categorization was applied to the reservoirs shown in Table 7.1 in order to produce the values in the last column. The colors in this table represent the aquifer categories shown in Figure 7.7. Notice that different reservoirs within a given field can be connected to aquifers of different strengths.

7.2.1 Aquifer impact on estimated ultimate recovery (EUR)

The available data on EUR per well for several GoM wells was used to arrive at a general probability distribution curve. Reservoirs were separated according to aquifer type into three categories, i.e. reservoirs with (a) depletion and/or poor aquifer drives, (b) intermediate aquifer drives, and (c) fully connected strong aquifer drives with wide well spacing. The impact of aquifer characteristics on well EUR is clearly reflected in the results graph shown in Figure 7.8.

Table 7.1 Aquifer strengths of typical GoM fields (as of January 2019)

	Reservoir	Aq/HC ratio	Aquifer influx, MMSTB	Oil production, MMSTB	Water production, MMSTB	Water injection, MMSTB	VRR, fraction	Depletion, psi	Time, years	Aquifer Index, rB/d/psi
Field A	A1	40-60	184	219	15	25	0.65	4,400	10	11.4
	A2	5-10	7	26	2	0	0.15	3,200	11	0.5
	A3	250	3.5	3	1	0	0.82	2,700	7.3	0.5
	A4	0.5	low	26.5	2.1	31.5	0		7.5	0
	A5	3	low	11	0.4	3.5	0		9.5	0
Field B	East	30	100	80	60	0	0.65	2,000	15	9.1
	W/N	15	38	70			0.4	2,600	12	3.3
	South	8	12	34			0.3	400	12	6.8
	SW	40-60	125	168	6	31	0.73	2,000	11	15.6
Field C	N	40-60	216	141	75	0	0.86	1,600	10	37
	E	10-60	10	33	0.1	0	0.24	800	10	3.4

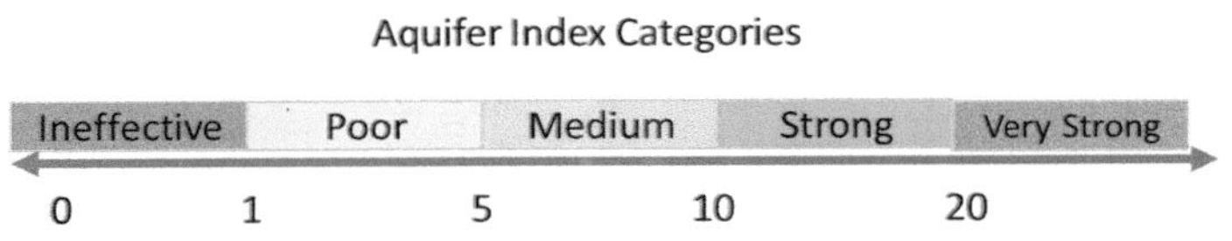

Aquifer Index Categories

Ineffective	Poor	Medium	Strong	Very Strong
0	1	5	10	20

Figure 7.7 Aquifer influx index categories.

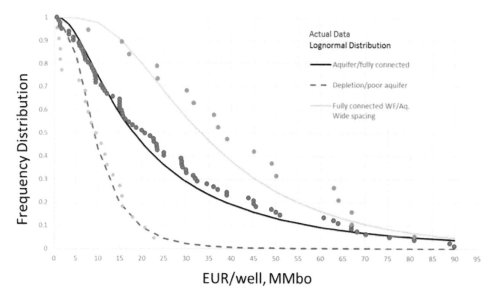

Figure 7.8 Estimated EUR/well for typical GoM reservoirs with weak, intermediate and strong aquifer drives.

7.2.2 Dynamic uncertainty of the aquifer

Because of uncertainty surrounding aquifer presence and strength, in many new field developments it is unclear whether water injection projects are needed. Due to such uncertainties, it is a good practice to conduct aquifer dynamic uncertainty analysis by including:

- intra-block transmissibility within the aquifer
- transmissibility between aquifer and reservoir, accounting for uncertainties around transition zone connectivity

A Monte Carlo simulation can be conducted for different cases of a given reservoir model in order to assess the magnitude of aquifer impact on the future development of any field, informing decision-makers on the necessity of water injection.

Our results indicate that:

- Water influx is independent of permeability range, which only impacts aquifer response time.
- Increasing aquifer porosity essentially increases aquifer size and significantly impacts its effectiveness.
- Aquifer compressibility affects both the level and timing of reservoir support.
- Due to the above, the aquifer to HC pore value (HCPV) ratio (AQ:HC) is not the only indicator of aquifer effectiveness.

7.3 WATER INJECTION TIPPING POINT

It is important to understand whether an aquifer can sustain continuous production in a new field. The water injection start time is another important operational decision with significant economic implications in such fields. To arrive at a water injection tipping point, several reservoir simulation runs were carried out for the aquifer drive of different AQ:HC (2:1–64:1). We used wettability as a proxy for aquifer strength. The charted recovery factor (RF) versus HCPV influx results are shown in Figure 7.9 (Fassihi and Blangy, 2022). The three curves represent RF for oil-wet, water-wet and intermediate wettability cases. The appropriate relative permeability curves were used for these scenarios.

The numbers on each curve represent a unique AQ:HC, and the triangles show the point where the impact of the aquifer is felt on the oil production curve. Notice that independent of wettability, aquifers with AQ:HC < 8:1 follow the same trendline. For weak aquifers, the incremental RF for AQ:HC from 8:1 to 64:1 is only 7% (32%–25%). For stronger aquifers, the corresponding incremental RF is 32% (56%–25%). Thus, for weak aquifers, it is important to augment aquifer influx with additional water injection irrespective of aquifer size. For strong aquifers however, it is advisable to delay injection until the size of the aquifer is determined.

Another important consideration is the placement of the water injection well. Depending on the severity of the impact that the transition zone has on connectivity impairment or permeability reduction with depth within the aquifer, such injectors could have a better outcome if placed above the WOC in the oil leg. As far as permeability reduction with depth is concerned however, injectivity could be lower due to the relative permeability effect. But higher permeability in the oil leg might more than compensate for such reduction.

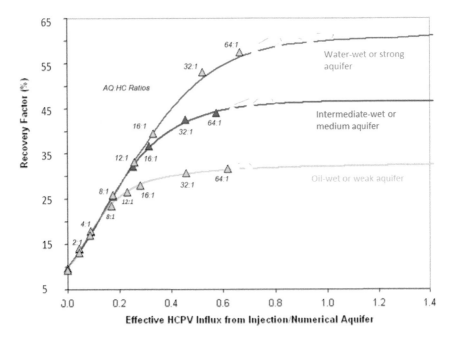

Figure 7.9 Water injection tipping point for a generic reservoir.

7.4 AQUIFER IMPACT ON ESTIMATION OF RESERVES AND RESOURCES

A key uncertainty in most oil reserve estimations is the type of drives in the reservoir. Generally, OOIP, rock compressibility and AQ:HC all act in the same direction, i.e. increasing RF. Unless there are strong indications of an aquifer's presence, connectivity and impact on oil recovery, the estimated reserves should only consider depletion drive performance.

Based on the authors' experience, we recommend incorporating and/or considering the following steps/issues when using simulation to book reserves for reservoirs expected to be impacted by aquifer influx:

- a thorough study of aquifer properties using geological and petrophysical data (size, permeability, porosity, rock compressibility, salinity, pressure)
- uncertainty about WOC (RFT sensitivity, etc.)
- impact of tar mats, if any present
- presence or absence of tilted aquifers and the physics of the tilt phenomenon
- impact of compartmentalization and faulting on aquifer influx
- impact of permeability degradation with depth on aquifer influx
- impact of initial overpressure (above bubble point) on aquifer flow
- uncertainty about aquifer response time (when will it kick in?)
- use of material balance to estimate aquifer strength and size
- sensitivity runs to aquifer size and its impact on RF using numerical models
- water injection support to aquifer and timing of injection
- well completion design (or use of deviated or horizontal wells) to control aquifers or prevent coning
- value of information for reducing aquifer uncertainty
- outrunning aquifer with increasing production rate in gas reservoirs (RF increase compared to volumetric depletion)
- changes in reserves with time due to aquifer strength unpredictability should be investigated as part of the development plan

7.5 CONCLUSIONS

- Because of the significant impact of aquifers on HC reservoir performance, they need to be adequately characterized using established geoscience and engineering modeling techniques.
- A newly defined parameter, AII, appropriately describes aquifer strength.
- In designing any future water injection projects, it is important to establish water injection tipping points considering all available aquifer and reservoir parameters.

REFERENCES

Ahmed, T. (2006). *Reservoir Engineering Handbook*. 3rd edn. Cambridge, MA: Elsevier.
Carter, R.D. and Tracy, G.W. (1960). An improved method for calculating water influx. *Petroleum Transactions of the AIME* 219 (1): 415–417. SPE-1626-G. Available from https://doi.org/10.2118/1626-G.

Coats, K.H. (1970). Mathematical methods for reservoir simulation. Presented by the College of Engineering, University of Texas at Austin, June 8–12.

Connolly, P. (2010). Robust workflows for seismic reservoir characterization. SEG Spring Distinguished Lecture. *Recorder: Official publication of the Canadian Society of Exploration Geophysicists* 35 (4): 7–10. https://seg.org/Education/SEG-on-Demand/id/6209/distinguished-lecture-recordings-robust-workflows-for-seismic-reservoir-characterization (accessed March 11, 2022).

Fassihi, M. R. and Blangy, J.P., 2022, Aquifer Influx Versus Water Injection in GoM. Paper presented at the SPE Improved Oil Recovery Conference, Virtual, April 2022. doi: https://doi.org/10.2118/209431-MS

Fetkovich, M.J. (1971). A simplified approach to water influx calculations-finite aquifer systems. *Journal of Petroleum Technology* 23 (7): 814–828. SPE-2603-PA. https://doi.org/10.2118/2603-PA.

Schilthuis, R.J. (1936). Active oil and reservoir energy. *Transactions of the AIME* 118 (1): 33–52. SPE-936033-G. Available from https://doi.org/10.2118/936033-G.

Van Everdingen, A.F. and Hurst, W. (1949). The application of the Laplace transformation to flow problems in reservoirs. *Journal of Petroleum Technology* 1 (12): 305–324. SPE-949305-G. Available from https://doi.org/10.2118/949305-G.

Chapter 8

Field case studies

INTRODUCTION

In order to illustrate the workflow for aquifer characterization discussed earlier in the book (see Section 7.1 and 7.2), in this chapter we present case studies for two fields in the Gulf of Mexico (GoM). We will discuss applications pertaining to both the static and the dynamic modeling aspects of the workflow. Although we have referenced field A elsewhere in the book (see Chapter 7), only fields B and C will be discussed in this chapter.

8.1 FIELD B: STATIC MODELING

Primary reservoirs in field B are sheet-like sands that have a very large areal extent. A relatively high sedimentation rate throughout the Miocene coupled with the creation of accommodation space due to high subsidence rates resulted in a significant burial depth, which, in turn, has exposed the reservoirs to higher temperatures. The temperatures and fluid movement over the highly permeable sheet sands initiated quartz overgrowths at the grain scale (thermal diagenesis) prior to the Pliocene, when most of the oil migration is thought to have occurred. Well data from field B indicates rock quality (RQ) (including storage and flow properties) degradation over the area. However, because well data does not include aquifer sampling, there is a paucity of regional analogue well data for aquifer properties. Therefore, forward modeling via Touchstone was used to predict rock quality in the aquifer portions of the reservoir system.

Seismic and well production data was first used to identify the presence of an aquifer system for both the northern and the southern parts of field B. Petrophysical data was then interpreted to estimate porosity and permeability. Figures 8.1 and 8.2 show the trends in porosity and permeability versus depth below mudline (DBML) for the three field cases (i.e. A, B and C).

Reservoir diagenesis was quantified using a two-step process. First, basin modeling was used to reconstruct the burial history, temperature and effective stress of field B (Figure 8.3). Second, Touchstone modeling was used to generate a quartz cementation trend are thought to have. RQ degrades more in the southern segment, where, according to modeling, temperatures reached 80 °C or more sooner (at approximately 10 million years) than in the northern flank. Porosity and permeability trends versus depth were predicted away from the well control, with average values for those rock properties calculated over the entire area.

158 Integrated Aquifer Characterization and Modeling

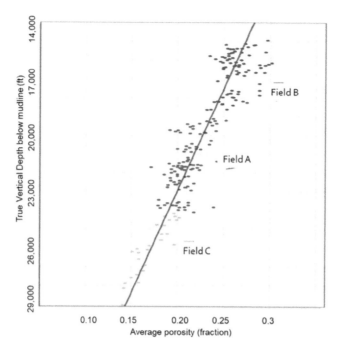

Figure 8.1 Porosity trend versus depth for fields A, B and C.

Supported by the subsurface interpretations above, the static model was built using the following two steps:

Step 1 *Assessing depositional trends*: porosity and permeability properties were populated with a facies constraint using a sequential Gaussian simulation (SGS) algorithm.

Step 2 *Quantifying diagenetic alteration*: porosity and permeability were reduced with depth by applying a factor through a mathematical function.

In addition to characterizing the amount and type of diagenesis, the workflow for modeling the aquifer system must also capture variations in other parameters, such as the individual segments' gross thickness, net-to-gross (NtG) and connectivity. Most large faults identified seismically were labeled as seals based on the integration of the interpreted offset along fault planes and on observed pressures from existing wells.

Table 8.1 shows the final aquifer properties.

8.2 FIELD B: DYNAMIC MODELING

Using the static model, the estimated OWIP was 3 bn bw. However, an engineering rationale advocated the necessity of incorporating additional numerical aquifer cells based on the aquifer's actual areal extent. Thus, 22 numeric aquifer grid sections (or legos) were set up for each sand area as shown in Figure 8.4.

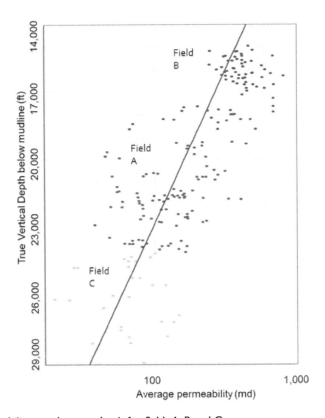

Figure 8.2 Permeability trend versus depth for fields A, B and C.

All grid blocks were connected using the non-neighbor connection (NNC) feature in the reservoir simulator. The lego-centroid DBML data was used to populate each aquifer grid block with the porosities and permeabilities predicted through Touchstone.

As discussed in Chapter 6, pressure buildup data is one of the key types of data for estimating aquifer transmissibility to the HC reservoir. For field B, one of the wells had a long shut-in period (see Figure 6.7). Monte-Carlo simulations were run in order to arrive at a good history-match for the pressure buildup and its derivative. However, in order to history-match the reservoir simulation to the production data, Touchstone properties had to be severely degraded. The properties adopted in the final simulation results are shown in Figure 8.5.

The final aquifer legos are shown within the dotted lines in Figure 8.6. The AQ:HC ratio was estimated at approximately 12, which is a smaller value than that predicted by the static model. The reason for this smaller value of the history-matched numerical aquifer as compared to the areal extent of the actual aquifer used in the dynamic simulation is due to presence of sub-seismic faults that tend to baffle the aquifer's energy and impact its connectivity.

160 Integrated Aquifer Characterization and Modeling

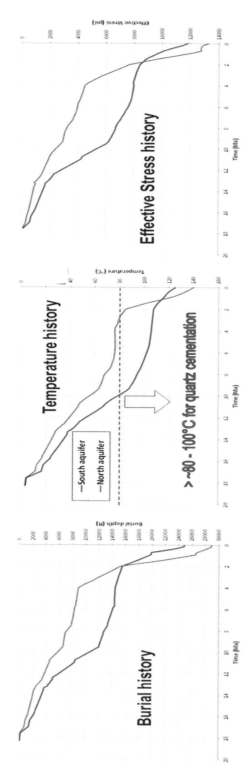

Figure 8.3 Estimated burial, temperature and effective stress histories for field B.

Table 8.1 Deterministic aquifer assumptions

Parameter	Low	Mid	High
Porosity, %	11	14	21
Permeability, md	1	6	235
Connectivity	Strong baffles	Mid baffles	No baffles
Gross TVT	Well average	>20% well average	>40% well average
AQ:HC ratio	3.9	8.4	16

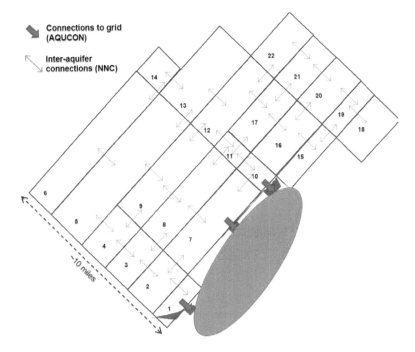

Figure 8.4 Aquifer grid sections (legos) for field B.

8.3 FIELD C: STATIC MODELING

Field C is a mature GoM field that has been in production for 15 years. It is a large, faulted anticline producing from mid-Miocene turbidite reservoirs. The reservoir consists of areally extensive amalgamated channel and/or layered sheet sands with good permeability and porosity containing undersaturated black oil.

A workflow similar to the one used for field B (see Chapter 7) was employed to characterize the aquifer in field C. We used EOD maps as well as horizon surfaces interpreted from a merged or "quilted" seismic dataset as input for the modeling. We assigned seismically interpreted faults displaying significant offset as barriers to flow and designated certain areas of the field as carrying a likelihood for presenting deformation bands and acting as baffles to flow. Given the extensive and rather continuous nature of the reservoir sands, the structural features (faults and deformation bands) were interpreted as controlling features defining the

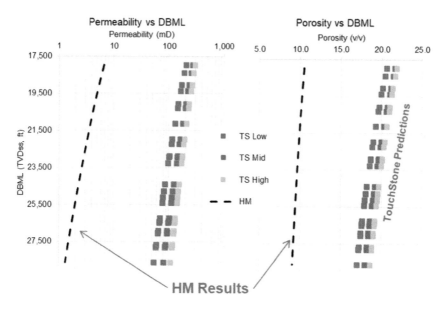

Figure 8.5 Final history-matched porosity-permeability trend.

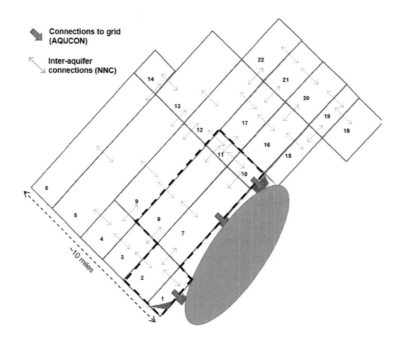

Figure 8.6 Final numerical aquifer (12:1) for field B.

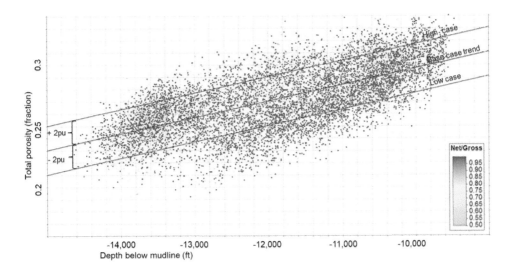

Figure 8.7 Porosity trend versus depth and uncertainty estimation (+/- 2 porosity units, pu) for Field C.

limits to the aquifer's extent. We interpreted low, base and high case true vertical thickness (TVT) maps for each interval of interest, leveraging the EOD maps as well as data from other nearby fields for guidance. Using the base reservoir and the top reservoir interpreted seismically from wells in the field, we derived a thickness map and created two simple grids representing the top and the base of the sands. The grids were populated with porosity and permeability properties using petrophysical relationships derived from well data extrapolated and applied to the aquifer. As a final step, we ran the uncertainty analysis tool (in Petrel) for estimating aquifer pore volume (PV) with varying input values for the area's variables such as GRV, NtG and porosity. As described in the volumetric assessment workflows of Chapters 4 and 2, we obtained P10/P50/P90 volumetric scenarios from approximately 200 stochastic simulations.

The areal extent of the aquifer (low, mid and high) was estimated using seismic combined with pressure data from the producing wells. The gross thickness was estimated for low, mid and high cases.

A simple grid was used for the area of interest and average aquifer properties were estimated as a function of depth using Touchstone modeling. A porosity trend was first established as a DBML function (Figure 8.7) and was then combined with a porosity-permeability relationship in order to arrive at an understanding of the range in permeability versus depth to be used as input for the dynamic model.

8.4 FIELD C: DYNAMIC MODELING

The aquifer size distribution from the regional aquifer study discussed above was used as input for the Eclipse simulation and the uncertainty analysis. Initial material balance calculations revealed an AQ:HC of 59. Large aquifer volumes were represented by numerical aquifers in the Eclipse simulation. The low case was defined by a 5 md cutoff which was applied to the

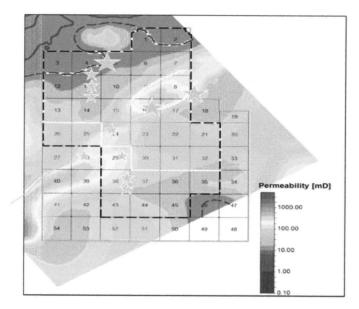

Figure 8.8 Low (white dash), mid (black dash) and high (orange dash) aquifer cases for Field C.

permeability map. Mid and high permeabilities, i.e. >5 md cutoff line, were outside the area of interest.

Using Monte-Carlo simulations and varying the aquifer parameters yielded good history-matches for well rates, well pressures and for observed water–oil ratios. This methodology provided the final three deterministic aquifer cases that were used to arrive at the estimation range for the HC reserves in the field (Figure 8.8).

Chapter 9

Applying petroleum lessons to aquifers during the energy transition

INTRODUCTION

In light of rising interest in carbon capture and sequestration (CCS), saline aquifers (SA) and depleted hydrocarbon reservoirs (DHR) are promising CO_2 storage sites for the future. Storing gas in aquifers is not a new concept: gas utility companies have been using aquifers to store natural gas as part of peak shaving efforts, meaning the proactive management of overall supply and demand. Moreover, recent interest in hydrogen fuel and air storage have created a need for underground hydrogen storage (UHS) or compressed air storage in aquifers (CAESA). All aspects of aquifer characterization and modeling discussed in previous chapters are applicable to gas storage or CCS, so we expect that petroleum geoscientists and engineers will play a key role here as well. However, injecting gas into aquifers faces some geomechanical and geochemical challenges that will be discussed in this chapter.

9.1 CO_2 STORAGE

Most world energy demand predictions identify hydrocarbon fuels as a key ingredient of the near-term mix of energy sources. However, these fuels' increased burning will result in more CO_2 emissions needing to be sequestered in order to meet global climate goals (net carbon zero by 2050). The International Energy Agency (IEA) estimates a 20% reduction of CO_2 emissions with the help of CCS within the next 30 years. Fulfilling this estimate may require capturing and storing up to 20 gigatons (Gt) per year of anthropogenic CO_2 from the atmosphere starting from 2050 onward (IPCC 2018). This is a dramatic increase from the current CCS capacity of 35 megatons (Mt) per year.

The objective of underground storage is to contain CO_2 for long durations in order to mitigate climate change. The first stage of CCS involves removing CO_2 from a flue gas through pre-, post- or oxyfuel combustion. The second stage consists in storing it safely for long periods of time. Currently, depleted oil and gas fields and deep saline aquifers are the most likely storage sites for sequestered CO_2 due to their sizable storage capacity on the one hand, and, on the other hand, because the infrastructure necessary for CO_2 injection is partly in place thanks to oil and gas production practices. Basalt formations are another potential long-term storage site based on the ability of CO_2 to form stable carbonate minerals following high-temperature reactions. For the time being, conventional reservoirs are the most viable storage site candidates because they afford a relatively greater amount of connected geological storage space than other unconventional reservoirs.

There have been various phases of technology development for carbon capture during the last century. The first phase of carbon capture started in the 1930s with the invention of the amine process to remove CO_2 from flue gas. The second phase took off in the 1970s for utilizing CO_2 as an enhanced oil recovery (EOR) process. The third phase began in the 1990s with the storage of CO_2 in geological formations (Herzog 2018). The world's first industrial-scale CO_2 injection project designed specifically as a greenhouse gas mitigation measure was developed in 1996 at the Sleipner field in the North Sea, about 240 km off the coast of Norway and operated by Statoil and the Sleipner partners. The CO_2 extracted from the natural gas (containing about 9% CO_2) produced on the offshore platform has been injected at a rate of 1 million tons per year into Utsira Sand, a major saline aquifer of late Cenozoic age, starting in October 1996 (Chadwick et al. 2008). The incremental investment for CCS was around $80 million. Under a CO_2 tax of $50/ton of CO_2, the payback would take only 1.6 years (Herzog 2018).

According to the databank of CO2RE, there are 73 CCS projects in development and currently in operation (CO2RE 2020). In the oil sector, three key CCS projects besides Sleipner are:

- In Salah Gas project (Algeria) started operations in 2004 and was subsequently suspended due to pressure rise in the formation following the injection of 3.8 million metric tons of CO_2.
- Snohvit project (Barents Sea) started operations in 2008 and injects 700,000 metric tons of CO_2 per year into a sandstone formation 2.6 km below the seabed.
- Gorgon project (Barrow Island, on the Northwestern Australian shelf), currently still inactive due to technical difficulties, was projected to capture 4 million tons of CO_2 per year from the produced gas (containing about 14% CO_2) and inject it into a sandstone formation 2.5 km below sea level.

In the industrial sector, there are two key CCS projects (Herzog 2018):

- Quest project (Fort Saskatchewan, Alberta, Canada) upgrades bitumen by adding hydrogen produced through a process called "steam methane reforming." CO_2, the byproduct of this process, is subsequently injected at a rate of 1 million tons per year into a 2 km deep saline formation 64 km away.
- Decatur project (Illinois, United States) produces fuel-grade ethanol through corn fermentation, a process that also generates CO_2 as a byproduct. The resulting CO_2 is injected at a rate of 1 million tons per year into a 2 km deep formation as of 2011.

In the power sector, there are two main projects (Herzog 2018):

- Boundary Dam project (Estevan, Saskatchewan, Canada) captures post-combustion CO_2 from a lignite coal power plant. The low cost of lignite and the size of the reservoir (300 years of supply) incentivized CCS use: the captured CO_2 is being injected at a rate of 1 million tons per year two km deep into sandstone since 2014. CCS use enabled this project to obtain a 20% government subsidy.
- Petra Nova project (Texas, United States) uses the CO_2 byproduct of a post-combustion pulverized coal power plant at a rate of 1.6 Mt CO_2 per year in EOR applications associated with the production of 240 MWe since 2016.

Moreover, hydrogen production from natural gas via ammonia is a growing industry that needs to dispose of large amounts of CO_2. Some of the new hydrogen players are planning their own CCS developments, such as the Pale Blue Dot in Acorn (Kjolhamar et al. 2021). Going forward, saline formations are prime targets for CCS application due largely to their global abundance and sizable storage capacity. In the United States, permitting of CO_2 injection wells entails the permitting of class-VI disposal wells, a relatively new class of wells requiring substantial amounts of integrated subsurface (e.g. geoscience and engineering) studies.

Potential storage for sedimentary basins around the world is estimated conservatively at around 8,200 Gt CO_2 and between 20,000 and 35,000 Gt CO_2 in high estimates (Kelemen et al. 2019). A reasonable CO_2 footprint for sequestration in sedimentary reservoirs lies between 0.5 and 5.0 t CO_2/m^2, and depends on the architecture of the reservoir and the caprock, the petrophysical properties of the rocks, the pressure and the temperature in the reservoir, and the extent of secondary trapping mechanisms (National Academies of Sciences Engineering Medicine 2019). Comparing these values with annual global CO_2 emissions highlights the potential these storage options have of significantly reducing CO_2 emissions to the atmosphere. The US Department of Energy (DoE) estimates for the prospective carbon storage sites in North America are provided in Table 9.1. An important takeaway is that saline formations could provide the biggest share of the future CO_2 storage capacity.

As of April 2018, the National Energy Technology Laboratory (NETL) database listed a total of 305 CCS projects (including 299 identified sites) worldwide (www.netl.doe.gov/coal/carbon-storage/worldwide-ccs-database). The 299 in situ projects include 76 for capture, 76 for storage, and 147 for both capture and storage in more than thirty countries across six continents. While some of the projects are either still in development or have already been completed, 37 are actively capturing and/or injecting CO_2.

Rising interest in the use of saline aquifers as CO_2 storage sites as part of CO_2 sequestration and storage practices has prompted researchers to identify the right conditions for optimum CO_2 storage. Given the fact that the storage potential of saline formations dwarfs other storage site options, NETL has mapped the largest aquifers and has created atlases of potential CO_2 sequestration sites. Figure 9.1 displays the mapped aquifers that could be used for CCS purposes. A comprehensive review of technical requirements for CO_2 storage is provided in the 2005 Intergovernmental Panel on Climate Change (IPCC) report.

Table 9.1 Storage resource estimates for North America

	CO^2 storage resource estimates (billion metric tons)		
	Low	Medium	High
Oil and natural gas reservoirs	186	205	232
Unmineable coal	54	80	113
Saline formations	2,379	8,328	21,633
Total	2,618	8,613	21,978

Source: DOE, data current as of November, 2014 (https://www.netl.doe.gov/node/5841).

Figure 9.1 Saline aquifers in North America. (Modified from Friedman et al. 2015).

9.1.1 Geological considerations

According to the American Petroleum Institute (API), the best CO_2 storage site options are salt formations, depleted oil fields and saline aquifers (depleted or not) (see Figure 9.2).

- Depleted oil fields are good candidates that provide economic incentive: injecting CO_2 into oil reservoirs has the added benefit of increasing short-term oil recovery.
- Salt formations (such as salt domes etc.) are ideal due to their excellent seal properties and highly efficient storage affordance.
- Saline aquifers, as discussed in the previous section, are both abundant and capacious storage sites. They are by far the most significant option in terms of afforded storage volume.

Aquifer characterization in view of assessing its suitability as a CO_2 storage site can be modeled on subsurface workflows used in oil and gas, as discussed in previous chapters. The key selection criteria for subsurface CO_2 storage reservoirs are the following:

- capacity (how much CO_2 can be stored)
- injectivity (the number of wells required impacts the cost of storage)
- containment (the ability to keep CO_2 underground)

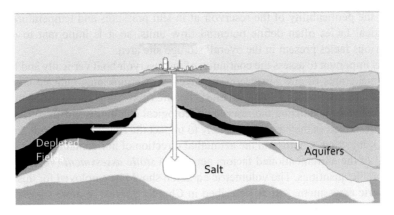

Figure 9.2 The three main types of CCS reservoirs: depleted fields, salt formations and saline aquifers. (Based on an image from www.energyinfrastructure.org/energy-101/natural-gas-storage.)

Storage site requirements depend greatly on the trapping mechanism and on subsurface geological conditions. Storage site characterization standards are currently under way and will be published in the near future (see Frailey et al. 2018). The Society of Petroleum Engineers (SPE), having previously developed the Petroleum Resources Management System (PRMS) and the attending classification of Petroleum Reserves and Resources (2007, 2011 and 2018), has again taken the lead in establishing standard industry guidelines and is presently working on the Storage Resources Management System (SRMS), with a projected publication date for the end of 2021. The International Organization for Standardization (ISO) has issued some standards and best practices pertaining to the geological storage of CO_2. Some examples are ISO 27918 Carbon dioxide storage using enhanced oil recovery (CO2-EOR), ISO 27914 Geological storage, and ISO 27923 Injection operations, infrastructure and monitoring.

In sum, geological considerations should include characterization of both the reservoir and the seal(s). There are different levels of rigor that are required, depending on the stage of a storage project.

- For *site screening* work, the analysis does not require wells.
- However, for the purpose of *site characterization*, and later of *site selection*, the work must include extensive calibration of the assessed reservoir and seal/containment properties to field data, which includes information from wells drilled in the subsurface.

9.1.1.1 The reservoir

- Reservoir characterization starts with the description of the depositional environment of the geological formation under consideration. Next, as is the case in the oil and gas industry, potential reservoir sites or storage segments and their estimated expanse must be identified.
- Once the reservoir is identified, its potential quality—generally proportional to the net reservoir thickness multiplied by its porosity—must be assessed.

- Injectivity is another important element of reservoir assessment, and it is determined by the permeability of the reservoir at in situ pressures and temperatures. The geological facies often define potential flow units, so it is important to examine the various facies present in the overall storage site area.
- It is important to assess the continuity of the reservoir both vertically and laterally, and to make note of any anisotropy displayed. Anisotropy may indicate the presence of faults and/or fractures. Such structural features may bring into question the possibility of compartmentalizing the reservoir. Geological continuity and anisotropy must both be described as they are highly likely to impact the spacing of the well, the size of the CO_2 injection pattern, and the azimuthal direction of horizontal injection wells.
- All of the aforementioned factors generate a **static assessment** (volume in place) of storable quantities. The volumetric equation should be employed for this initial volumetric assessment, just as described in Chapter 2. Next, a more rigorous geomodel building project should be developed that can serve as the framework for a flow/storage simulation.
- Geological descriptions impact the **dynamic model** that simulates CO_2 storage in the reservoir. For example, in situ geological layering and the attending depth-dependent permeability gradient can serve to control vertical injection conformance. Similarly, horizontal injection conformance may be impacted by lateral heterogeneity in the geological facies.
- In the case of CO_2 injection in a legacy oilfield for EOR applications, the flow behavior involves multi-phase fluids (oil, water, gas, CO_2). Given the relatively higher permeability rate of CO_2 as compared to that of oil or water, it is therefore important to evaluate the presence and size of the bottom aquifer.

9.1.1.2 The seal(s)

In addition to the description and characterization of the aquifer, the evaluation of overburden properties is of paramount importance during screening. Overburden properties include (Chadwick et al. 2008):

- The stratigraphy of the seal (specifically the various lithologies and their thicknesses).
- The nature of any faults and/or fractures present in the seals that border the storage container and throughout the overburden, from seal to surface.

If present, shallower aquifers near the surface can be sampled and monitored for the early detection of upward CO_2 migration from the storage site. The characterization of the overburden will be discussed further in sections 9.1.3 and 9.1.4.

9.1.1.3 Data requirements

The following data types are generally useful for aquifer characterization (IPCC 2005: 225):

- Raw data
  - seismic profiles across the area of interest, preferably three-dimensional or closely spaced two-dimensional surveys
  - wells with subsurface sampling (data details below)

- Container geometry
    - structure contour maps of reservoirs, seals and aquifers
    - detailed maps of the structural boundaries of the trap where CO_2 is to accumulate, including highlighted potential spill points and lateral limits
    - integrated container edge (ICE) maps
- Reservoir characterization, continuity and compartments
    - documentation and maps of faults
    - maps of the predicted CO_2 migration pathway from the point of injection
    - facies maps showing all lateral facies variation in the reservoirs or seals
    - samples of core and drill cuttings from the reservoir and seal intervals
    - well logs (preferably a consistent suite), including geological, geophysical and engineering logs
    - calibration of log-derived properties to core data
    - petrophysical measurements, including porosity, permeability, mineralogy (petrography), seal capacity, pressure, temperature, salinity and laboratory rock strength testing
- Characterization of fluids
    - fluid analyses and tests from downhole sampling and production testing
    - oil and gas production data (for hydrocarbon fields)
    - pressure transient tests measuring reservoir and seal permeability
    - special core analysis (SCAL)
    - pressure, temperature, water salinity; in situ stress analysis determining (a) fault reactivation potential and fault slip tendency, and (b) maximum sustainable pore fluid pressure during injection vis-à-vis reservoir, seal and faults
- Fluid flow expectations
    - hydrodynamic analysis identifying the magnitude and direction of water flow, the hydraulic interconnectivity of formations, and the pressure decrease associated with hydrocarbon production
- State of present-day stresses
    - seismological, geomorphological and tectonic data indicating neotectonic activity
    - passive seismic monitoring (microseismicity)

9.1.1.4 Data availability and cost

Even though common for oil and gas field development and operations, obtaining data for the purpose of characterizing brine aquifers often requires special justification due to high cost. The importance of complete pre-existing hydrocarbon exploration and production datasets cannot be overemphasized, because these datasets contain crucial information for characterizing the subsurface. For instance, no seismic data exists if target areas are not already designated as potential hydrocarbon or geothermal resource sites. Yet without pre-screening potential sites for seismic potential, little to no structural control is available away from wells. Moreover, fewer wells are drilled for the specific purpose of site characterization alone, and, consequently, less data is generated to calibrate subsurface properties with reasonable certainty.

When candidate storage sites are located outside of existing hydrocarbon basins, additional funds are required so as to obtain the data needed to characterize saline aquifers. Chadwick et al. (2008) recommend the following datasets for a robust characterization of disposal reservoirs and their overburden:

- a regular grid of two-dimensional (2D) seismic data over an area sufficiently broad to characterize overall reservoir structure
- a high quality three-dimensional (3D) seismic volume of the injection site and adjacent area, tuned, if possible, for a satisfactory resolution of both reservoir and overburden
- sufficient well data to allow for reservoir and overburden characterization

Key geological indicators for screening potential storage sites are provided in Table 9.2 (adapted from Chadwick et al. 2008):

In conclusion, among the geological limitations to potential storage sites are internal compartmentalization and flow barriers (such as complex hydrogeologic heterogeneity due to lithology or faults), unexpected diagenesis-induced change in porosity and/or permeability resulting in limited injectivity, and induced or triggered seismicity. The viability of all candidate projects is clearly determined by an assessment of the amount of capital required relative to the benefits—including government incentives—derived.

9.1.1.5 The importance of economic analysis

Economic screening is important for selecting potential storage sites. Lateral compartmentalization helps retain CO_2 in the desired storage site, but it can also impair CO_2 injectivity and require subsequent elevated injection pressures (Chadwick et al. 2008), increasing costs. To justify the investment (which includes the cost of CO_2 capture, transportation and injection) and to reduce the cost per ton of sequestered CO_2, the best sites will usually exhibit greater storage capacity and longer sequestration lifespan. Accordingly, storage sites that include reservoirs with individual CO_2 sequestration capacity larger than 1 Mt each are good candidates for initial consideration.

9.1.1.6 Resources available for screening analysis

A tool developed by the US DoE's NETL—CO2 Storage prospeCtive Resource Estimation Excel aNalysis (CO2-SCREEN)—can screen saline formations for prospective CO_2 storage resources. A dependable method for calculating prospective CO_2 storage resources, CO2-SCREEN also provides consistency across result comparisons between different research efforts. CO2-SCREEN consists of a Microsoft Excel spreadsheet of geologic data linked to a Monte-Carlo simulation model that calculates prospective CO_2 storage resources (www.youtube.com/watch?v=lhakk-HYfOI).

9.1.2 Fluid properties

Reviewing the physical properties of carbon dioxide and water is critical for understanding the role of aquifers as sites for CO_2 storage. Figures 9.3 and 9.4 show the viscosities and densities of CO_2 and water as a function of temperature and pressure.

Table 9.2 Key geological indicators for screening potential storage sites

	Positive indicators	Cautionary indicators
Reservoir efficacy		
Static storage capacity	Estimated effective storage capacity significantly larger than total amount of CO_2 to be injected	Estimated effective storage capacity close to total amount of CO_2 to be injected
Dynamic storage capacity	Predicted injection-induced pressures well below levels likely to induce geomechanical damage to reservoir or caprock	Injection-induced pressures approaching geomechanical instability limits
Reservoir properties		
Depth, m	>1000, <2500	<800, >2500
Net reservoir thickness, m	>50	<20
Porosity, %	>20	<10
Permeability, mD	>500	<200
Salinity, g/l	>100, to avoid spoiling potable water resources	<30
Reservoir complexity (compartmentalization): stratigraphic and structural	Uniform stratigraphy, lack of faults or fractures	Complex lateral variation and complex connectivity of reservoir facies. Potential structural compartmentalization (faults and/or fractures)
Caprock efficacy		
Lateral continuity	Stratigraphically uniform, with small or no faults	Lateral variations, medium to large faults
Thickness, m	>100	<20
Capillary entry pressure	Much greater than maximum predicted injection-induced pressure increase	Similar to maximum predicted injection-induced pressure increase

Notice that CO_2 viscosity can be correlated to CO_2 density as shown in Figure 9.4b below. CO_2 viscosity can be estimated using the equation shown in this figure.

Interestingly, on the one hand brine viscosity is strongly dependent on temperature, whereas CO_2 is not. On the other hand, CO_2 density is more dependent on the pressure to temperature ratio than brine. In most screening studies, CO_2 assumes a density of 800 kilograms (kg) per cubic meter (m^3). Figure 9.5 shows CO_2 volume relative to depth. CO_2 density increases rapidly at a depth of approximately 800 meters (m), when CO_2 reaches a supercritical state—"above critical point," i.e. T>31.1 °C, P>72.9 atmospheric pressure (atm) or 1,072 pounds per square inch (psi). Such contrasting physical properties strongly impact the migration of CO_2 plume in the aquifer. Because of its lower density, CO_2 tends to rise and is driven by buoyancy to accumulate at the highest possible place in the reservoir, beneath the seal. The relative volume CO_2 occupies (represented by bubbles) decreases dramatically as depth increases. At depths below 1.5 km, CO_2 density and specific volume become nearly constant.

CO_2 is ideally stored in a supercritical state, or at a corresponding depth of 800 m or 2,600 ft below the Earth's surface, when it acquires the density of a liquid while maintaining the

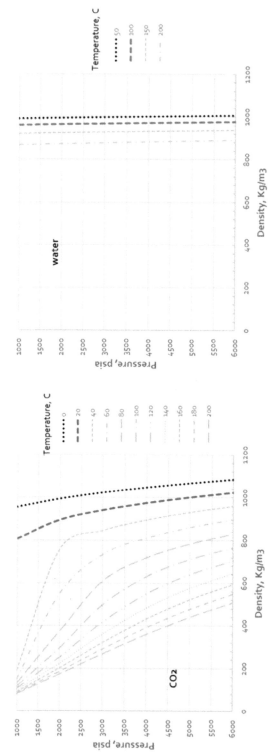

Figure 9.3 CO_2 and water density versus temperature and pressure. (Data from National Institute of Standards and Technology.)

Applying petroleum lessons to aquifers 175

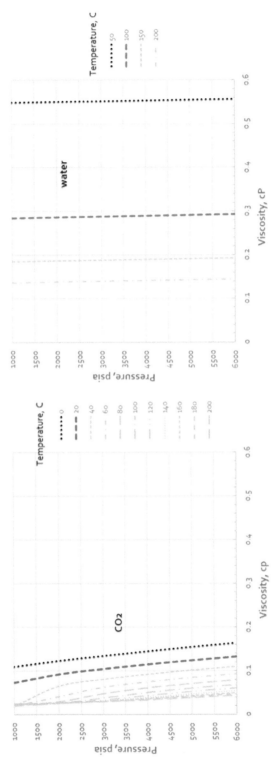

Figure 9.4a CO_2 and water viscosity versus temperature and pressure. (Data from National Institute of Standards and Technology.)

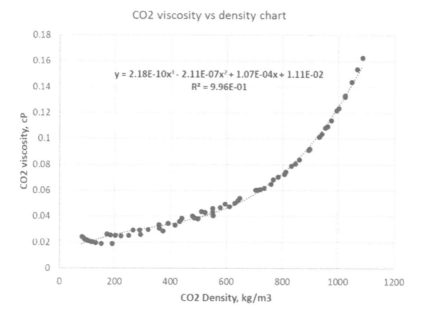

Figure 9.4b CO_2 viscosity vs. density.

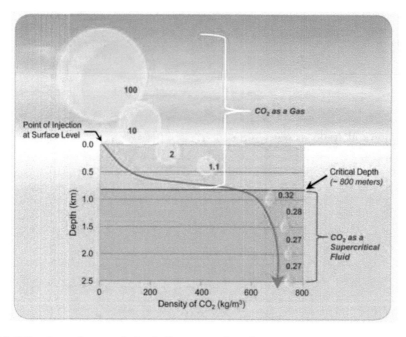

Figure 9.5 CO_2 volume change with depth. (For a normal geothermal gradient and hydrostatic pressure gradient) (reproduced with permission from Friedman et al. 2015: 6.)

viscosity of a gas, because in this state the largest CO_2 amount possible can fit in a given reservoir volume.

CO_2 solubility in brine is dependent on several factors, such as pressure, temperature and salinity (Figure 9.6). A plot of CO_2 solubility relative to depth for both fresh water and brine (35,000 ppm salinity) is provided in Figure 9.7. A temperature gradient of 25 °C per km and a hydraulic gradient of 0.433 psi/ft for fresh water and 0.441 psi/ft for saline water were assumed in these plots. Notice that CO_2 solubility decreases with increasing salinity, as expected. Similarly, solubility increases sharply up to a depth of 800 m, after which its rate of increase slows down.

The use of saline aquifers as sites for CO_2 storage requires a deep understanding of salinity and its effects on CO_2 aggregation. Because high-salinity formations lower CO_2 solubility, or the ability of the gas to dissolve into the brine, less saline formations are likely to be in higher demand. They can be identified by using the data resulting from existing technologies for well logging, such as downward spectroscopy, which measures formation chlorine. Coupled with other conventional logs, this data can give a robust, continuous estimation of aquifer salinity. Other remote sensing techniques—the controlled source electromagnetic (CSEM) method, electrical resistivity, etc.—may be useful as well, as discussed in Chapter 2.

During the injection phase, spectroscopy-logging techniques can be used to measure CO_2 saturation. One should keep in mind that CO_2 dissolution in brine is a slow process, entailing a time scale of hundreds to thousands of years.

Figure 9.8 compares isothermal compressibility and sonic velocities at 23 °C for pure water, CO_2 and methane. Based on this data, different tools could potentially be used to detect water, methane or CO_2 displacement in subsurface rocks (e.g. water shows the highest sonic velocity whilst methane shows the highest compressibility). Unfortunately, the percentage of dissolved CO_2 cannot be measured accurately with seismic data due to the small rate of change in water compressibility. However, we expect that the plume of concentrated CO_2 is detectable in the reservoir, since pure CO_2 and water have significant differences in compressibility and primary wave or pressure wave (P-wave) velocity. Moreover, CO_2 should be detectable if it migrates upward to shallower zones (more on this topic in section 9.1.4).

9.1.3 CO_2 containment modeling

CO_2 containment in storage over circa 1,000 years is the ultimate goal of any CCS project. There are many ways that CO_2 can escape the storage container. The main typical classes of CO_2 migration are displayed in Figure 9.9.

Reservoir simulation and geomechanical studies are needed to assess the risk associated with different migration classes. The proper data on relative permeability must be used to properly model the CO_2 plume migration. During the injection phase of a CO_2 storage project, the movement of CO_2 is dominated by drainage relative permeability, as CO_2 displaces brine, i.e. the formation's wetting phase. After injection, the governing force on the injected fluid is gravity. Under most reservoir conditions, CO_2 is more buoyant than brine and, as a result, injected CO_2 tends to migrate up-dip along the top seal. Over hundreds of years, the migration of the CO_2 plume is dominated by both drainage and imbibition relative permeability effects. At the head of the migrating plume, drainage relative permeability is prominent as water drains away from the migrating plume. At the tail of the migrating plume, imbibition relative permeability dominates as water imbibes behind the migrating plume

178 Integrated Aquifer Characterization and Modeling

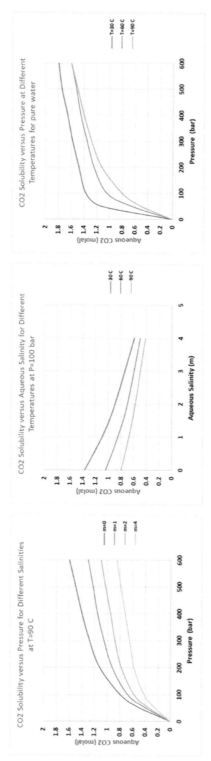

Figure 9.6 CO_2 solubility as a function of temperature, pressure and salinity. (Based on data from Spycher and Pruess 2004.)

Applying petroleum lessons to aquifers 179

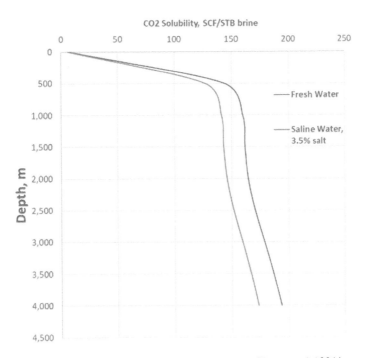

Figure 9.7 CO_2 solubility versus depth. (Based on correlations by Chang et al. 1996.)

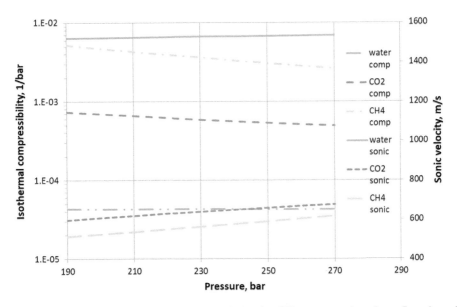

Figure 9.8 Isothermal compressibility and sonic velocity for CO_2, water and methane (based on data from http://webbook.nist.gov/chemistry/fluid).

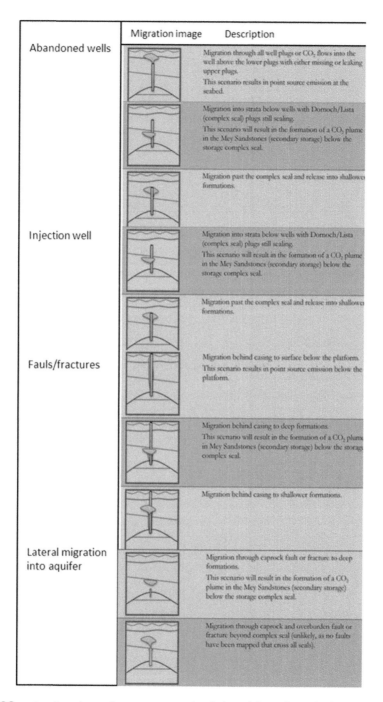

Figure 9.9 CO_2 migration classes from a storage site (adapted from Scottish Carbon Capture and Storage Group 2015).

trapping CO_2 as a residual phase. Thus, the amount of mobile CO_2 diminishes over time, as more of the plume dissolves and is trapped as a residual phase (Flett et al. 2004).

In the absence of hard data, assessing storage capacity and injectivity can rely on the use of analogs. In order to account for uncertainty in such measurements however, value ranges should be established as inputs for the following:

- gas relative permeability at initial water saturation
- trapped gas saturation to brine
- water relative permeability at trapped gas saturation

9.1.4 Geomechanical modeling requirements

Geomechanical risks during CO_2 injection into an aquifer are associated with the following:

- CO_2/brine/rock interactions
- tensile failure of the reservoir
- shear failure of the reservoir
- tensile failure of the caprock
- shear failure of the caprock
- thermal fracturing near the injector
- thermal fracturing of the caprock
- fault slip

Because CO_2 may leak through the caprock and sealing horizons, their integrity is paramount to the qualification of a potential storage site (IPCC 2005: 227). Whenever fluid is injected into a porous and permeable reservoir rock, injection pressure needs to exceed formation pressure at the time of injection. In a petroleum reservoir, formation pressure at the time of injection is expected to be lower than the formation's initial pressure on account of depletion. It is thus possible to safely inject the fluids. In an aquifer however, bottomhole injection pressure exceeds the initial formation pressure. This pressure can generate stress in the reservoir rock or in the seal rock, causing existing fractures to reopen or creating new fault and/or fracture planes.

Geomechanical modeling studies can help assess the maximum formation pressure sustainable in situ for any given storage site (IPCC 2005: 227). Such studies require data on pore fluid composition, mineralogy, in situ stresses, pore fluid pressures as well as general characterizations of the overburden, including pre-existing fault orientations and their frictional properties. The application of this methodology at a regional scale is documented by Gibson-Poole et al. (2002). Mini-frac tests on lower perm seals should provide a good estimate of seal permeability. Moreover, due to the buoyancy of a CO_2 column relative to brine, capillary pressure measurements are needed to assess the capillary entry pressure for the seals. Capillary entry pressure should be compared with the pore pressure increase caused by injection immediately below seals. As discussed by Chadwick et al. (2008), the integrity of the top seal is a prerequisite due to CO_2 buoyancy. It follows that when the reservoir or the aquifer is dipping, the nature of the lateral seal becomes important.

As discussed in Chapter 3, the efficacy of an oil or gas field's top seal, i.e. its potential ability to hold a certain column height, is determined by the seal's capillary entry pressure. If

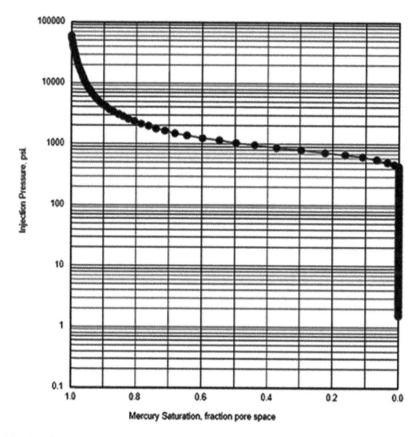

Figure 9.10 Capillary pressure for a typical top seal.

we substitute the latter with CO_2 buoyancy forces against the seal, the same reasoning applies to an aquifer's top seal. Figure 9.10 plots capillary pressure measurements for a top seal in the case of a mercury injection. The entry pressure for this sample is around 400 psi. The maximum height of the CO_2 column under this caprock is estimated at 4,000 ft as follows:

$$\Delta P = 0.433 \, (\rho_{brine} - \rho_{CO2}).h \qquad (9.1)$$

Use $\rho_{brine} = 1.08$, $\rho_{CO2} = 0.85$ \qquad (9.2)

h = 4,000 ft

9.1.5 Geochemical modeling requirements

Another important part of the characterization of a saline aquifer as a CO_2 storage site consists in the geochemical profile of interactions between CO_2, brine and rocks (IPCC 2005: 227). The mixture of CO_2 and brine generates dissolved CO_2, carbonic acid and bicarbonate ions.

Once dissolved into the fluid formation, CO_2 has the following reaction equations (Thomas et al. 2016):

$$CO_2(g) \leftrightarrow CO_2(aq) \qquad (9.3)$$

$$CO_2(aq) + H_2O(l) \leftrightarrow H_2CO_3(aq) \qquad (9.4)$$

$$H_2CO_3(aq) \leftrightarrow H^+(aq) + HCO^-(aq) \qquad (9.5)$$

$$HCO_3^-(aq) \leftrightarrow H^+(aq) + (CO_3)^{2-}(aq) \qquad (9.6)$$

The dissociation of carbonic acid into active hydrogen ions and bicarbonate ions (equation 9.5) may trigger complex reactions with geological fluids and formation rocks, immobilizing in the aqueous and mineral phases. Pore water acidification reduces the amount of CO_2 that can be dissolved. Consequently, rocks that buffer the pore water hydrogen ion concentration (pH) to higher values (reducing acidity) facilitate the storage of CO_2 as a dissolved phase. Water rich in CO_2 is likely to react with minerals in the reservoir rock or the caprock matrix, or with the primary pore fluid. Such reactions cause fracture occlusion and sealing through the precipitation of carbonates along the walls of fractures or inside pores, thus impairing permeability.

A full assessment of the potential effects such reactions can have in systems with more complex mineralogies requires sophisticated simulations validated by the results of laboratory experiments conducted using reservoir and caprock samples and native pore fluids in settings recreating in situ conditions. We emphasize that predictions of the long-term impact of geochemical reactions cannot be directly validated on a field scale, because these reactions may take a long time. However, simulations of important mechanisms—such as the convective mixing of dissolved CO_2—can be tested through comparisons to laboratory analogues (IPCC 2005: 228).

Because of the gravity override of the CO_2 plume, which is due to its unfavorable mobility ratio, the plume might move under the original oil–water contact (OWC) during injection phase. Once injection ceases, the viscous driving force is removed, leaving only the buoyancy force driving CO_2 up-dip and back above the site of original contact.

9.1.6 Monitoring requirements

Some of the key possible environmental impacts due to injection of CO_2 in a reservoir are:

- Leakage-related events (Chadwick et al. 2008)
    - groundwater pollution from migrating CO_2 causing dissolution and alteration of minerals from rocks and soils, potentially contaminating fresh water supplies
    - CO_2 surface leaks forming accumulations in local depressions and confined spaces hazardous to humans and other living organisms
    - CO_2 leakage impacting the biodiversity of ecosystems
- Induced seismicity
    - CO_2 injection, like the injection of disposal water, may trigger induced earthquakes, depending on the state of stress in the area.

In the oil and gas industry, many studies have documented that the re-injection of large amounts of waste water can modify in situ stresses to the point of inducing seismicity, depending on the state of local stresses. When the pumping stops, or when it is reduced below a certain trigger limit, microseismic events also stop.

Up until recently there had been no documented cases of CO_2 injection leading to induced seismicity, but the geophysics of the Earth's stress and strain is the same as in waste water injection. Dichiarante et al. (2021) recently published a study from the Decatur CCS site, where CO_2 has been injected from 2011–2014 and from 2017 to the present at the base of the Mount Simon sandstone saline reservoir. Injection was cautiously undertaken as deep basement faults had been identified from reprocessed seismic data. Injection pressure at the site was held significantly below fracture pressure; nonetheless, induced seismic events occurred and were observed to spread beyond the expected extent of the CO_2 plume. The study identified microseismic clusters and bursts as well as unresolved weaknesses on a smaller scale, and argued that local stress transfers related to the injection of CO_2 reactivated pre-existing fractures in the critically stressed basement, thus inducing seismicity.

Proper characterization in the early phase of an injection project and monitoring of the aquifer in the later stages should mitigate such risks to the environment. Simulations of the long-term CO_2 migration rate, direction, and rate of dissolution in the formation are critical for the design of cost-effective monitoring programs. Simulation results are the basis for the technical design of storage sites' monitoring aspects, such as the location of monitoring wells and the frequency of remote sensing surveillance away from wells and of soil gas or water chemistry sampling. Moreover, once they are calibrated through history-matching observations of injection and monitoring operations, simulation models serve to test multiple scenarios and to assess the impact of possible operational changes. Examples might include the decision to drill new wells or to alter injection rates, with the goal of avoiding CO_2 migration past a likely spill-point (IPCC 2005: 228). If in situ stresses at the site of injection indicate that induced or triggered earthquakes may occur, it is recommended to monitor microseismicity with a surveillance array of geophones placed either in wellbores, at the surface, or in both locations.

Note that the time scale for CO_2 to dissolve in the aqueous phase is typically short when compared to the duration of the CO_2 migration out of the storage formation by means of other processes. Once dissolved, CO_2 can eventually be transported out of the injection site through basin-scale circulation or upward migration. However, the time scale—millions of years—for such transport is typically so long that the eventuality of such transport can arguably be omitted from leakage risk assessments.

Generally speaking, the objectives of a monitoring program are to:

- Ensure and document injection well conditions, and to measure injection rates as well as wellhead and formation pressures. Based on the experience of the petroleum industry, injection well leakage resulting from improper completion or from the deterioration of the casing, packers or cement is often the lead cause of failure for injection projects.
- Verify the quantity of injected CO_2 stored through various mechanisms.
- Optimize storage project efficiency, including storage volume use, injection pressure determinations, and drilling rationale for new injection wells.

- Deploy the appropriate techniques to demonstrate that CO_2 is contained in the intended storage formation(s). This is currently the main method of assuring that CO_2 remains in storage and that performance predictions are verified.
- Detect leakage and provide early mitigation plan warnings.

In terms of surveillance, key monitoring tools used for "quantification and verification" include seismic, electromagnetic (EM), electrical methods, distributed acoustic sensing (DAS) and distributed temperature sensors (DTS). As discussed in Chapter 2, some of these methods can be expensive. A new, more cost-effective monitoring technology is being developed that relies on deploying sparse DAS arrays on the seafloor (in either active or passive mode) for offshore storage sites. Other emerging technologies relevant for future monitoring practices involve four-dimensional (4D, or time-lapse) tools that help monitor pressure fronts, saturation plumes and zonal stress changes.

9.1.6.1 A 4D example

As discussed in Chapter 3, 4D seismic data relies on measured differences in seismic amplitude and travel time between baseline and repeat monitor surveys. 4D monitoring registers detectable changes in the aggregate compressibility of the rock and fluid correlated with changes in CO_2 saturation. As a general rule, more compressible rocks (e.g. unconsolidated sands or turbidites) produce better 4D signals than stiffer rocks (such as carbonates). It should be noted that 4D signal strength depends on the noise level, which in turn is a function of the frequency of seismic surveys and the quality of the seismic processing.

Sleipner, a large gas field in the North Sea's Norwegian sector with Paleocene and Jurassic sandstone reservoirs, was first developed in 1974. The overlying Miocene Utsira Formation was subsequently selected as a CO_2 storage site where 12.1 Mt of CO_2 were injected between 1996 and 2010 (White et al. 2018, Williams and Chadwick 2017). The growth of the CO_2 plume at the Sleipner project site is shown in Figure 9.11. The structural

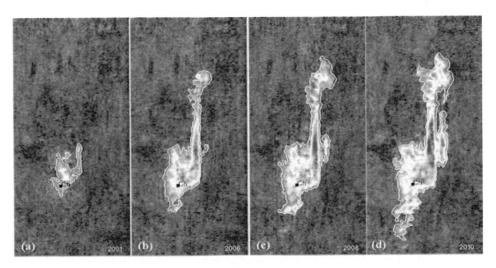

Figure 9.11 Growth map of the topmost CO_2 layer with time showing the growth of the CO_2 plume at the Sleipner project site. (Williams and Chadwick 2017.)

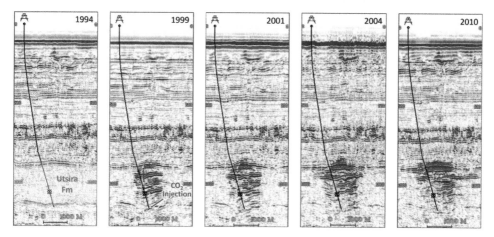

Figure 9.12 Vertical (depth) view of the changes in seismic with time due to injection of CO_2 in the Utsira formation at Sleipner. (Dupont et al. 2021.)

map at the base of the caprock (the top seal) suggests that the infill of CO_2 is driven by buoyancy. These maps provide important data on CO_2 conformance in the Utsira geological container at Sleipner.

As discussed by Dupont et al. (2021), 4D seismic data is also effective at tracking the vertical sweep of CO_2 in addition to the horizontal sweep described in Figure 9.11. As can be seen clearly in Figure 9.12, there are several layers in the Utsira package receiving CO_2, with the topmost layer showing the most 4D change through time. It should be noted that the interval receiving the CO_2 is capped by a seal of excellent quality barring surface access. 4D monitoring and verification has shown no detectable events above the seal so far. As discussed in Chapter 2, small events below seismic detection may be in the process of forming, which is why the project combines other techniques (surface sampling, geochemical analyses) to supplement 4D monitoring. Dupont et al. deployed several different methods in their analysis, and concluded that pre-stack seismic (AVO) data generated more accurate injection volume estimates than post-stack data. This is consistent with our discussion of seismic data types in Chapter 2.

Recent fiber-optic-based field trials have identified the latter as the best existing technology for leak detection and for monitoring seismicity. The CO_2 data detection limit from a storage site using this technology suggests it is able to identify minimal amounts of CO_2 (1.3–10.6 tons, 3.3–27.4% gas saturation) in the shallow subsurface. This is about 30 to 300 times smaller than the amounts of CO_2 that conventional seismic data can detect (Kjolhamar et al. 2021).

A summary of established techniques for monitoring CO_2 storage is provided in Table 9.3 (IPCC 2005: 236).

9.1.7 Sequestration capacity estimation

The main driving rationale for developing a geological CO_2 storage site involves estimating the potential volume of CO_2 that can be stored at that site and demonstrating that the site is capable of meeting required storage performance criteria. A good summary of different methodologies for storage capacity estimation is provided in the IEA report published in 2013.

Table 9.3 Summary of established direct and indirect techniques for monitoring CO_2 storage projects

Category	Technique	Measurement	Application
Sampling	Tracers (introduced and/or natural)	Travel time	Tracing movement of CO_2 in the storage formation
		Partitioning of CO_2 into brine or Oil	Quantifying solubility trapping
		Identification sources of CO_2	Tracing leakage
	Water composition	CO_2, HCO_3^-, CO_3^{2-}	Quantifying solubility and mineral trapping
		Major ions	Quantifying CO_2 water-rock interactions
		Trace elements	Detecting leakage into shallow groundwater aquifers
		Salinity	Detecting leakage into shallow groundwater aquifers
	Soil gas sampling	Soil gas composition	Detecting elevated levels of CO_2
		Isotope analysis of CO_2	Identifying source of elevated soil gas CO_2
			Evaluate impact on ecosystems
In situ measurements	Subsurface pressure	Formation pressure	Control of formation pressure below fracture gradient
		Annulus pressure	Wellbore and injection tubing condition
		Groundwater aquifer pressure	Leakage out of the storage formation
	Well logs	Brine salinity	Tracking CO_2 movement in and above storage formation
		Sonic velocity	Tracking migration of brine/CO_2 into shallow aquifer
		CO_2 saturation	Calibrating seismic velocities/ amplitudes for 3D seismic
Remote sensing	Time-lapse 3D seismic (4D)	P and S wave velocity	Tracking CO_2 movement in and above storage formation
		Reflection horizons & faults	Identification of container/seal in 3D; compartments
		Seismic amplitudes and/or Inversion	Identification of container/seal in 3D; compartments
		Seismic amplitude attenuation	Identification of container/seal in 3D; compartments
	Vertical seismic profiling (VSP)	P and S wave velocity	Detecting detailed distribution of CO2 in the storage formation
		Reflection horizons & faults	
	Distributed acoustic sensing (DAS), distributed temperature sensing (DTS)	High resolution seismic	Detection leakage through faults and fractures

(continued)

Table 9.3 Cont.

Category	Technique	Measurement	Application
	Cross-well imaging	Seismic amplitudes and/or Inversion	Detecting detailed distribution and/or leakage of CO2
		Seismic amplitude attenuation	Detecting detailed distribution and/or leakage of CO2
	Passive seismic monitoring	Location, magnitude and source characteristics of seismic events	Development of microfractures in formation or caprock Detect possible CO_2 migration pathways Establish and monitor state of stress in storage and caprock
Remote sensing with potential fields	Electrical and electromagnetic methods	Formation conductivity	Tracking CO_2 movement in and above storage formation
		Electromagnetic induction	Detect migration of brine/CO_2 movement in and above storage formation
	Time-lapse gravity measurements	Density changes caused by fluid displacement	Tracking CO_2 movement in and above storage formation CO_2 mass-balance in the subsurface
Remote sensing with potential fields	Land subsurface deformation	Tilt	Detect geomechanical effects on storage formation and caprock
		Vertical and horizontal displacement using interferometry and GPS	Locate CO_2 migration pathways
	Visible and infrared imaging from satellites or aircraft	Hyperspectral imaging of land Surface	Detect vegetative stress and/or changes in soil moisture
	CO_2 land surface flux monitoring using flux chambers of eddy variance	CO_2 fluxes between the land surface and atmosphere	Detect, locate and quantify CO_2 releases

Source: Adapted from IPCC (2005).

Storage capacity is assumed to be the total pore volume that can be occupied by injected CO_2. For oil reservoirs, this volume is analytically determined by estimating water saturation behind the CO_2 injection front using fractional flow and the Buckley-Leverett solution (Dake, 1978). However, there are other factors that can increase sequestration capacity:

- structural and stratigraphic trapping
- dissolution of CO_2 into the formation brine
- residual CO_2 trapping
- chemical reactions of CO_2 with minerals present in the formation

The significance of the last three factoring processes grows with time, as is shown schematically in Figure 9.13. It should be noted that storage capacity assessment is complicated by the different temporal and spatial scales over which these processes occur.

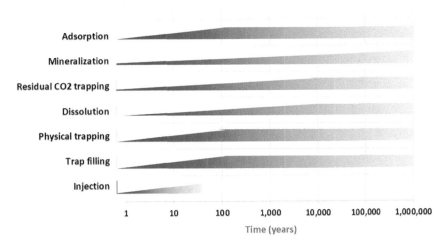

Figure 9.13 The time evolution of various CO_2 storage mechanisms operating in deep saline formations during and after injection. (Adapted from IPCC 2005.)

9.1.8 The role of aquifers for CO_2 sequestration in depleted reservoirs

As discussed by Bachu and Shaw (2003), the CO_2 storage capacity of a depleted gas reservoir, M_{CO2}, can be estimated from:

$$M_{CO2} = \rho_{CO2res} \cdot R_f \cdot (1 - F_{IG}) \cdot OGIP \cdot \frac{P_s Z_r T_r}{P_r Z_s T_s} \tag{9.7}$$

where
  ρ_{CO2res} is CO_2 density at reservoir conditions;
  R_f is the recovery factor;
  F_{IG} is the fraction of injected gas;
  P, T and Z denote pressure, temperature, and the compressibility factor, respectively;
  Subscripts "r" and "s" denote reservoir and surface conditions, respectively.

The capacity of a depleted oil pool can be estimated from:

$$M_{CO2} = \rho_{CO2res} \cdot R_f \cdot A \cdot h \cdot \varnothing \cdot (1 - S_w) - V_{iw} + V_{pw} \tag{9.8}$$

where
  A and h are area and thickness, respectively;
  ϕ is porosity and $(1 - S_w)$ is oil saturation;
  V_{iw} is volume of injected and/or invading water;
  V_{pw} is volume of produced water.

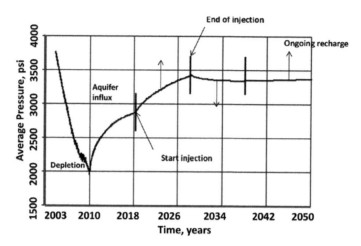

Figure 9.14 Reservoir pressure change due to depletion and subsequent CO_2 injection. (Adapted from Scottish Carbon Capture and Storage Group 2015.)

When oil reservoirs underlain by aquifers are produced, their pressure declines and aquifers invade the pore space that would have been available for CO_2 storage. The pressure equalization resulting from this depletion makes the injection of CO_2 more difficult. However, the latter depends on aquifer strength; weaker aquifers do not significantly impact CO_2 storage capacity in depleted reservoirs. Figure 9.14 shows the history-matched results and the forecast of average reservoir pressure for a depleted reservoir in offshore Scotland that was a candidate for CO_2 injection. We note that upon production cessation, aquifer influx raises reservoir pressure. Upon CO_2 injection, the pressure increases even more as expected but stays below the initial reservoir pressure. Following injection termination, ongoing recharge due to plume migration continues without significant pressure change.

Bachu and Shaw (2003) arrived at generalized criteria for establishing the strength and effect of underlying aquifers on CO_2 sequestration capacity in depleted oil and gas reservoirs in Alberta (see Table 9.4). They used the water–oil or the water–gas ratio as the primary criterion for establishing the strength of an underlying aquifer. The overall recovery factor provides a secondary criterion. The analysis of the Alberta pools with the largest CO_2 sequestration capacity confirmed that oil reservoirs with strong aquifer support and gas reservoirs with weak aquifer support have generally higher recovery factors and thus higher reduction in storage capacity. The corresponding coefficients of reduction in CO_2 sequestration capacity are shown in Table 9.4.

To arrive at the effective storage capacity, Bachu and Shaw suggest the following equation:

$$M_{CO2eff} = C_m . C_b . C_h . C_w . C_a . M_{CO2res} \tag{9.9}$$

where M_{CO2eff} is effective reservoir capacity for CO_2 sequestration, and the subscripts m, b, h, w and a stand for mobility, buoyancy, heterogeneity, water saturation and aquifer strength respectively. Some of these can be combined to arrive at an overall effective coefficient:

Table 9.4 Criteria for establishing the strength and effect of underlying aquifers on the CO_2 sequestration capacity of depleted oil and gas reservoirs in Alberta

Pool type	Aquifer strength	WOR range, WGR	Average RF, % (range)	CRC
Oil pools	strong	>= 0.2	38 (5–78)	0.40
	weak	<0.2	19 (10–40)	1.20
Gas pools	strong	>=10	72 (40–96)	0.72
	weak	<10	83 (30–93)	1.00

Source: Bachu and Shaw (2003).

Note: CRC = capacity reduction coefficient
RF = recovery factor
WOR = water–oil ratio (m^3/m^3)
WGR = water–gas ratio (bbl/MMcf)

$$C_{eff} = C_m . C_b . C_h . C_w \qquad (9.10)$$

Bachu and Shaw suggest that:

C_{eff} <0.3 for aquifers
C_{eff} =0.5 for depleted oil reservoirs
C_{eff} = 0.9 for depleted gas reservoirs

Chadwick et al. (2008) provide expressions for estimating the capacity factor, C, as shown below:

$$C = C_{gas} + C_{liq} \qquad (9.11)$$

$$C_{gas} = \langle \emptyset . S_g \rangle \qquad (9.12)$$

$$C_{liq} = \langle \emptyset . S_l . X_{lCO2} . \frac{\rho_l}{\rho_g} \rangle \qquad (9.13)$$

where
C is the fraction of the aquifer volume that can contain CO_2;
C_{gas} is free supercritical CO_2;
C_{liq} is CO_2 dissolved in the native pore fluid;
S_g and S_l are volume fractions of the pore space containing supercritical CO_2 and liquid respectively;
$X_{l\,CO2}$ is the mass fraction of CO_2 dissolved in the brine;
φ is the effective porosity;
ρ_g and ρ_l are the densities of the supercritical and liquid phases respectively;
the angle brackets represent averaging over the spatial domain of storage.

For a regional aquifer, Chadwick et al. (2008) propose the following expression:

$$Q = A.D.\varnothing.\rho_{CO_2}.h_{st} \tag{9.14}$$

where

Q is storage capacity (kg);
A is the areal distribution of the aquifer (m²);
D is the cumulative thickness of good reservoir rocks (m);
h_{st} is storage efficiency, i.e. fraction (by volume) of the reservoir pore space that can be filled by CO_2 (in free or dissolved form);
ρ_{CO2} is the density (kgm⁻³) of pure CO_2 at reservoir conditions.

In conclusion, estimating storage efficiency is highly uncertain as it depends on multiple parameters such as reservoir boundaries, heterogeneity, salinity, pressure, temperature, etc. In their review of several reservoirs in the North Sea, Chadwick et al. (2008) mention storage efficiency values in the range of 0.0009–0.4.

Physical trapping and dissolutions are considered the two most important factors in estimating storage volumes for aquifers. The following example illustrates a quick estimation of storage capacity for an aquifer located in the Gulf of Mexico:

Depth: 10,000 ft
Hydrostatic gradient: 0.433 psi/ft
Average temperature: 250 F
Aquifer volume: 10 billion barrels (bn bbl)
Salinity: 50,000 ppm
Average trapped gas saturation: 5%

Using Equations 9.3 through 9.6, one arrives at:

Estimated Trapped CO_2 = 0.045 Gt
Dissolved CO_2 = 0.084 Gt
Total sequestered CO_2 = 0.129 Gt

By equating this value with equation 9.4, h_{st} is estimated to be 0.14. This value may be much lower in reality due to limitations on injection pressure. The latter will have a lower impact as the size of the aquifer increases. An example of estimated storage efficiency in SA is shown in Figure 9.15. We note that the early sharp rise in efficiency is due to the free CO_2 phase following the injection phase. Later on, due to CO_2 plume dissolution in the saline aquifer and to chemical reactions, efficiency increases. Moreover, efficiency is much lower for bounded aquifers than for open boundary aquifers. In the former case, the low compressibility of brine and rock does not allow for a significant volume of CO_2 to be injected. In the latter case, plume migration allows for a much larger CO_2 volume to be injected.

When there is sufficient rock and fluid properties' data to characterize an aquifer in three-dimensionality (3D), numerical reservoir simulators are effective tools for testing injection strategies and for forming a reasonable estimate of the ultimate achievable storage in that

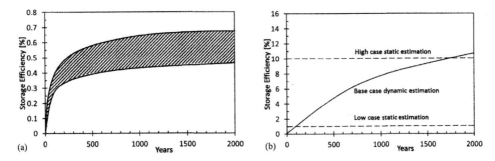

Figure 9.15 Estimated storage efficiency for (a) bounded and (b) unbounded SA. (Alkan et al. 2021.)

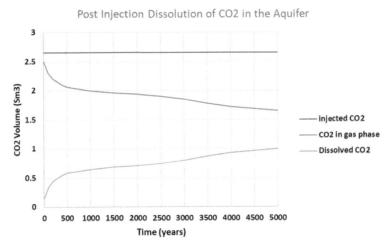

Figure 9.16 Post-injection dissolution of CO_2 in reservoir. (Redrawn based on Chadwick et al. 2008.)

aquifer. The standard simulators that are used for hydrocarbon production forecasting can be used for this purpose. Inclusion of geochemical and geomechanical features would greatly enhance the reliability of these simulators. For CO_2 stored in the liquid or supercritical phase at reservoir conditions, storage capacity is significantly better than storage capacity for CO_2 in gaseous phase. However, water displacement by CO_2 in a liquid state needs to be properly modeled.

An example of simulated CO_2 dissolution in the Froan Basin area is shown in Figure 9.16. The CO_2 phase change from free gas to dissolved gas in the reservoir is displayed on this graph. As the plume migrates slowly up the dip-slope, dissolution progressively reduces the volume of remaining free CO_2, with around 40% dissolved after 5,000 years. The CO_2 that is in solution sinks to the bottom of the reservoir where it will remain ponded in a gravitationally stable state. The long-term distribution of fluid following CO_2 injection termination causes the saturation of gas trapped within the aquifer to slowly heal itself through geochemical reactions.

9.1.9 Injectivity considerations

CO_2 injectivity depends not only on the properties of the reservoir rock itself, but also on the nature of its boundaries. If the aquifer is compartmentalized with flow baffles then very little CO_2 can be injected into it, as the only space available will be that created by the compression of the water and rock. Under these conditions, injection pressure increases and may come to exceed initial formation pressure.

Long-term storage volume depends on how much native fluids can be displaced from the reservoir over the duration of injection. For most aquifers, this displacement takes the form of groundwater migration into adjacent formations and/or to the ground or seabed surface. A CO_2 injection rate of 2 million tons per year over a period of 25 years is assumed in most screening studies. This corresponds to the typical lifetime output of a small power station.

Technologies for drilling, injection, stimulation and completion of CO_2 injection wells were established based on extensive oil industry experience using CO_2 injection for EOR. These technologies are being put in practice, with some adaptations, in current CO_2 storage projects.

The principal well design considerations for a CO_2 injection well include:

- pressure
- corrosion-resistant materials
- production and injection rates

The number of wells required for a storage project will depend on several factors, including total injection rate, permeability and thickness of the formation, maximum injection pressures and availability of land-surface area for the injection wells. For the most part, fewer wells will be needed for high-permeability sediments in thick storage formations and for projects with horizontal injectors. Horizontal and extended reach wells are good options for improving the CO_2 injection rate of individual vertical wells. For CO_2 injection through existing and old wells, key considerations include the mechanical integrity of the well, the quality of the cement behind the casing, and the state of well maintenance.

As one might expect based on experience from injection wells in oil and gas fields, maximum injection pressure is a key operational parameter that should not exceed fracture initiation and/or fracture propagation pressures in the formation. The US Environmental Protection Agency (EPA) Underground Injection Control Program (UIC) has shown that fracture pressures can range from 11 to 21 kilopascals (kPa) per meter (0.486 to 0.928 psi/ft) (IPCC 2005: 233).

Well drilling and well completion integrity (for both injection wells and abandoned wells in a reservoir that come in contact with the CO_2 plume) are very important for preventing surface leaks.

We briefly mentioned economic considerations earlier in this chapter. It stands to reason that the project must either be financially sustainable or receive incentives in order to be sanctioned. Chadwick et al. (2008: 50) state that:

> In CO_2 storage exploration, potential profitability is likely to be low and, moreover, the real-world integrity of an aquifer storage site may be difficult to test, even when an exploration or potential injection well that has been drilled. This is likely to put off investors unless the balance of risk and reward changes.

9.2 NATURAL GAS STORAGE

Storing natural gas in subsurface storage sites during periods of low demand and withdrawing it from storage when demand is at a high peak is a proven strategy for maximizing profits. According to the Energy Infrastructure Council (2016), there are approximately 400 active storage facilities in 30 different states within the US:

- Approximately 20 percent of all natural gas consumed during the five-month winter season each year is supplied by underground storage.
- There are three principal types of underground storage sites in use in the US today, i.e. depleted natural gas or oil fields (80%), aquifers (10%) and salt formations (10%).
- The capacity associated with operating underground natural gas storage in the US has increased by 18.2 percent between 2002 and 2014, helping ensure that natural gas is available when it is most needed.
- Approximately 4 trillion cubic feet (Tcf) of natural gas can be stored or withdrawn for consumer use.

Data from the US Energy Information Administration (2022) indicates that total natural gas storage capacity in the US in 2019 was 9.2 Tcf, of which 1.4 Tcf was aquifer storage capacity.

Qian et al. (2004) studied aquifer use for storing associated gas produced from deep-water fields for future recovery. This application is different from the seasonal storage practices of utility companies and, in fact, even though it concerns methane (CH_4) sequestration, it is in some ways similar to the newer application of CO_2 storage.

9.2.1 Methane properties

The corresponding graphs for CH_4 solubility in brine are provided in Figures 9.17 through 9.19. These figures are based on the proposed correlations by Spivey et al. (2004).

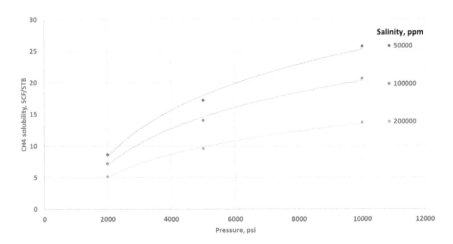

Figure 9.17 CH_4 solubility as a function of pressure and salinity at 225°F.

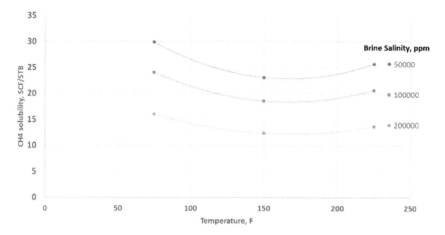

Figure 9.18 CH$_4$ solubility as a function of temperature and salinity at 10,000 psi.

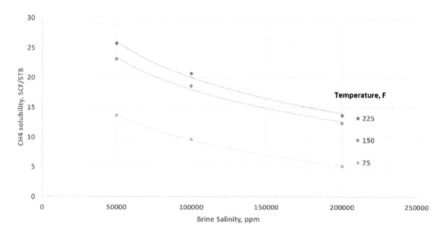

Figure 9.19 CH$_4$ solubility as a function of salinity and temperature at 10,000 psi.

9.2.2 Gas trapping

A key aspect of gas injection in saline aquifers is the gas-trapping mechanism, which we discuss at length in Chapter 3. When gas is injected into an aquifer, water is displaced by gas and water saturation is reduced until a critical gas saturation (S_g) is reached at which the gas begins to flow. As S_g increases, the relative permeability of gas also increases: gas enters the largest pores first and then invades progressively smaller pores. When gas injection stops and upon production, the direction of saturation change is reversed from decreasing (drainage) to increasing (imbition) water saturation. Water imbition into the smaller pores causes gas trapping in larger pores (Qian et al. 2004).

Figure 9.20 shows this hysteresis effect using drainage and imbition relative permeability curves (Core Lab data). In this case gas saturation reaches 66%, indicating the maximum

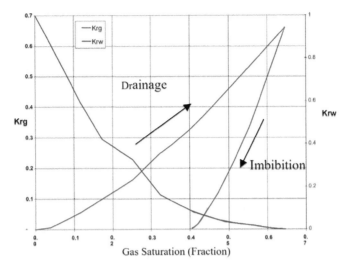

Figure 9.20 Gas-water relative permeability with hysteresis. (Core Lab data.)

pore space available as potential gas storage. Upon flow reversal, 40% of the pore space will retain the injected gas (trapped gas). Thus, about 60% of the total gas (40/66= .606) is trapped. A theoretical maximum percentage of recovered gas in this case is roughly 39% ((66-40)/66=.394). For areas of the reservoir that do not reach the maximum value of 66 percent gas saturation, the trapped gas percentage is even higher.

This analysis fails to take into account the impact of gas dissolution in water. According to its results, gas storage attempts are therefore likely to result in substantial loss of the injected gas, which would remain trapped in the aquifer. Qian et al. (2004) noticed that a steeper aquifer dip equates to a significant improvement of storage efficiency due to gravity segregation and better sweep.

An important factor in assessing storage efficiency is the magnitude of the residual (trapped) gas saturation. This value depends on the rock type and can be determined accurately with special core tests for a given aquifer. Although low recovery efficiency suggests aquifers are not a viable gas storage option, the same cannot be said about aquifer use for CO_2 sequestration. Indeed, due to high trap efficiency, aquifers may in fact be very suitable for this latter purpose.

In the case of natural gas storage in aquifers, compartmentalized reservoirs are preferred because they contain the injected gas and prevent it from migrating. However, in consequence, more injection wells are needed in order to fill up the reservoir, which provides operators with a more efficient and flexible storage operation. Bulk gas saturation in natural gas storage sites can be more than 50 percent by volume (Chadwick et al. 2008).

9.2.3 Best practices for natural gas storage in aquifers

Aquifers may be suitable for natural gas storage if the water-bearing sedimentary rock formation is overlain with an impermeable cap rock, and granted that they do not form part of, meaning that they are separated from, the drinking water aquifers. The key aspect of risk mitigation in any underground storage project is well integrity, and this is also the case for CO_2 storage.

Requirements for methane injection wells also apply to drilling petroleum wells (see Energy Infrastructure Council 2016), and are similar to those used for CO_2 injection. An operator's site-specific risk assessment provides guidance for decision-making on casing/tubing requirements for the design of new wells as well as for existing well completions.

9.3 UNDERGROUND HYDROGEN STORAGE (UHS)

Hydrogen injection into aquifers as part of future energy storage practices and the installation of offshore wind turbines are two areas that are being researched intensively, and results are slowly coming out. As a no-carbon energy carrier, hydrogen may play a significant role in the energy transition phase. For the most part, characterizing an aquifer for hydrogen storage follows the aforementioned workflow. However, there are several hydrogen-specific questions (Mouli-Castillo et al. 2021):

- What are the potential biological and chemical reactions between hydrogen and the rock, caprock, well cements/casing and fluids?
- What are the flow and trapping mechanisms of hydrogen in the aquifer, including any contamination, pore blocking or permeability reduction?
- How much hydrogen can be stored and what is the recovery efficiency?
- How would the public react to the idea of underground hydrogen storage and its potential environmental risks?

Unlike CO_2 storage, hydrogen trapping and dissolution result in loss of recoverable hydrogen, and, hence, they are not desirable. Similar to CO_2, hydrogen solubility increases with pressure and decreases with salinity. Academic research programs at institutions of higher education, such as the University of Edinburgh, are actively studying these questions. For example, a caprock good at sealing natural gas may leak if used to seal hydrogen due to its lighter density and much smaller molecule size.

Yekta et al. (2018) presented the results of their experimental measurements of capillary pressure and relative permeability for hydrogen–water fluid mixtures in a Triassic sandstone with 19% porosity and 44 mD permeability. They indicated that for potential hydrogen storage pressures (< 100 bar) and temperatures (< 100 °C), because hydrogen viscosity does not change significantly (~9 microPa.s), the capillary number, and therefore the relative permeability in the hydrogen–water system, will not be largely modified upon changing pressure and temperature in these ranges. This is different from a CO_2–water system, where capillary numbers can strongly vary with pressure and temperature. The measured relative permeability data by Yekta et al. (2018) are displayed in Figure 9.21. More measurements are needed before definitive correlations are recommended.

Lysyy et al. (2022) carried out pore-scale studies using microfluidics to study multi-phase hydrogen flow in an aquifer storage scenario. Their measured static and dynamic contact angles ranged from 17° to 56°, which confirms the non-wetting hydrogen nature. Because of the dissolution of hydrogen in aquifers, they suggested using cushion gases with low solubility in water. Since UHS is an emerging field, the numerical simulators that are currently being used for estimating UHS feasibility need to be validated with hydrogen laboratory data.

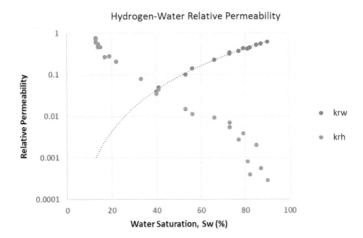

Figure 9.21 Hydrogen–water relative permeability. (Based on data provided by Yekta et al. 2018). Experiments were conducted at 20 and 45 °C and 55 and 100 bar. Trend lines are drawn by the authors of this book.

Current estimated costs of hydrogen storage range from 2 to 15 US dollars per kilowatt-hour (KWh). It is expected that hydrogen will eventually become cost-competitive with other alternative energy resources (Mouli-Castillo et al. 2021).

9.4 COMPRESSED AIR ENERGY STORAGE

Compressed air energy storage in aquifers (CAESA) can be considered a novel and potential large-scale energy storage technology in the future. The data for this clean energy production technology are currently scarce and many researchers are pursuing this field.

The compressed air energy storage system consists mainly of the surface energy storage power station and the gas storage device. The gas storage device includes the surface gas storage tank and the underground gas storage reservoirs which are mainly underground salt caves, hard rock caverns and aquifers. A typical CAESA layout is presented in Figure 9.22.

Similar to any other gas storage in porous rock, the caprock integrity is paramount in selecting a site for air storage to prevent any air leakage. As discussed by Zhang et al. (2019), the key factors to consider for reservoirs are tectonic history, sedimentary environment, horizontal continuity, thickness, porosity, frac pressure, permeability, mineral composition, fracture density and rock characteristics. These factors should be investigated in detail when selecting a CAESA site. Generally, frac pressure and permeability of the caprock are the main laboratory measurements to be carried out in the early phases of the project. A good caprock should have very low porosity and permeability and have a high frac pressure.

An ideal aquifer for air storage should have a high permeability for higher air deliverability. However, a bounded aquifer that prevents migration of injected air is preferred. Figure 9.23 presents the air solubility data for pure and saline water (0-6 salt molality, mole salt/kg of water) at different pressures and temperatures (based on data from Zhang et al. 2019).

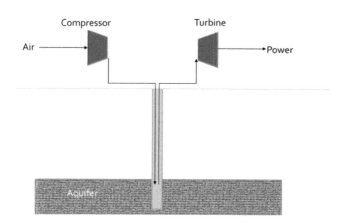

Figure 9.22 Schematic of a compressed air energy storage in aquifer systems.

The presence of oxygen in the injected air could lead to occurrence of some oxidation reactions in the aquifer resulting in a decrease in the oxygen content of the produced air. This could affect the combustion efficiency of the gas turbine and reduce power generation efficiency of the whole system (Zhang et al. 2019). The effect of oxygen reduction on the system may be more significant when the time scale is longer.

Other features of CAESA are designing the right wellbore structure that can handle both oxygen and high pressure. The impact of shock due to cyclic injection and production in the wellbore should be included in the selection of reasonable wellbore structure, wellbore material, drilling, completion and cementing methods (Zhang et al. 2019).

9.5 WASTE LIQUID DISPOSAL BY INJECTION INTO NON-POTABLE AQUIFERS

Injecting industrial waste liquids into confined underground saline aquifers is a good disposal alternative from both an environmental and an economic standpoint. In fact, injecting liquid waste (hazardous or non-hazardous) into deep aquifers is an ongoing practice as old as mining and/or drilling. The production of oil and gas from unconventional oil & gas reservoirs (e.g. shale) has generated knowledge useful to the topic of liquid waste disposal. Among the key takeaways relevant in this context is knowledge pertaining to the precautionary measures that need to be implemented before and during the disposal of significant volumes of produced water:

- The industry now knows that the characterization of in situ stress and its monitoring through injection are important for preventing and/or mitigating induced seismicity. Multiple field studies have shown that, under certain conditions, microseismic events can be triggered by the injection of waste liquid which stop once injection ceases.
- One of the key selection criteria for disposal sites is the projected impact of injection on the aquifer system and its equilibrium in the region.

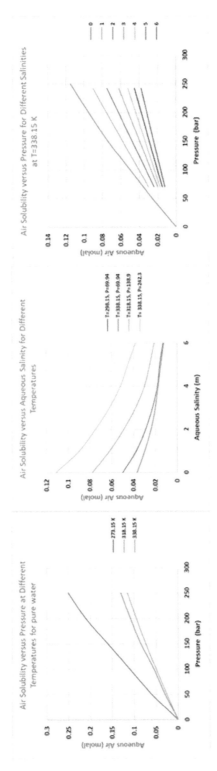

Figure 9.23 Air solubility in water at different pressures, temperatures, and salinities. (Based on data from Zheng et al. 2019.)

Aquifer characterization for waste disposal follows the same procedure as discussed for other applications. The geomechanical aspects of such operations (e.g. top seal integrity, fault reactivation, and migration to other connecting aquifers) require further research. Reservoir simulation of waste flow in the aquifer can be done using 3D transient mathematical finite-difference models that incorporate fluid properties such as density and viscosity as functions of pressure, temperature and composition. Such models are generally single-phase flow, allowing for both energy and mass transport by convection and conduction.

The EPA has established specific regulations for injection wells, and underground injection exemptions for some aquifers (www.epa.gov/uic). Aquifer exemptions allow groundwater sources to be used by energy, mining and other companies for the purpose of resource extraction and/or waste disposal purposes, provided such activities are in compliance with the EPA's UIC requirements under the Safe Drinking Water Act (SDWA).

9.6 CONCLUSIONS

Aquifer use for gas storage (CO_2, CH_4, air, etc.) or waste disposal entails the types of data analysis and research methods discussed in this chapter. Any forecasting effort needs to take into consideration much more than container size: the geomechanics and the geochemistry of injected CO_2 or air and their interaction with the formation rock and fluid are of critical importance for investment decisions and, beyond, for the integrity and sustainability of project operations in general. The ongoing monitoring of developed storage sites is also essential to verify, through time, the model predictions around containment in the subsurface. While government regulations can impose standards or timelines on such surveillance, robust reservoir/seal modeling with properly defined geosystems can outstrip concerns related to such restrictions by providing insights into any potential leaks over time.

We emphasize that all aspects of aquifer characterization related to petroleum reservoirs are applicable to the energy transition. Therefore, petroleum geoscientists and engineers are well positioned to tackle CCS initiatives or gas storage well into the future.

REFERENCES

Alkan, H. et al. (2021). Engineering design of CO_2 storage in saline aquifers and depleted hydrocarbon reservoirs: Similarities and differences. *First Break* 39 (6): 69–81. https://doi.org/10.3997/1365-2397.fb2021047.

Bachu, S. and Shaw, J. (2003). Evaluation of the CO_2 sequestration capacity in Alberta's oil and gas reservoirs at depletion and the effect of underlying aquifers. *Canadian Journal of Petroleum Technology* 42 (9): 51–61.

Chadwick, A. et al. (2008). Best Practice for the Storage of CO_2 in Saline Aquifers: Observations and Guidelines from the SACS and CO2STORE Projects. British Geological Survey Occasional Publication. Amersham: Halstan.

Chang, Y., Coats, B.K. and Nolen, J.S. (1996). A compositional model for CO_2 floods including CO_2 solubility in water. *SPE Reservoir Evaluation & Engineering* 1 (2): 155–160. SPE-35164-PA. https://doi.org/10.2118/35164-PA.

CO2RE (2020). Homepage. https://www.globalccsinstitute.com.

Dake, L.P., 1978, *Fundamentals of Reservoir Engineering*, New York: Elsevier Science Publishers.

Dichiarante, A.M. et al. (2021). Identifying geological structures through microseismic cluster and burst analyses complementing active seismic interpretation. Tectonophysics 820: article 229107. Available from https://doi.org/10.1016/j.tecto.2021.229107.

Dupont, M.A., Philit, S. and Cubizolle, F. (2021). A straightforward workflow for monitoring CO_2 storage. *GeoExpro* 18 (5). www.geoexpro.com/articles/2021/10/a-straightforward-workflow-for-monitoring-co2-storage.

Energy Infrastructure Council (2016). Underground natural gas storage (webpage). www.energyinfrastructure.org/energy-101/natural-gas-storage.

Flett, M., Gurton, R. and Taggart, I. (2004). The function of gas-water relative permeability hysteresis in the sequestration of carbon dioxide in saline formations. Paper presented at the SPE Asia Pacific Oil and Gas Conference and Exhibition, October 2004, Perth, Australia, SPE-88485-MS. https://doi.org/10.2118/88485-MS.

Frailey, S.M., Koperna, G.J. and Tucker, O. (2018). The CO_2 storage resources management system (SRMS): Toward a common approach to classifying, categorizing, and quantifying storage resources. Paper presented at the 14th International Conference on Greenhouse Gas Control Technologies (GHGT-14), October 21–26, Australia.

Friedman, J.S. et al. (2015). The Carbon Storage Atlas. 5th edn. DoE report by the National Energy Technology Laboratory.

Gibson-Poole, C.M., Lang, S.C., Streit, J.E., Kraishan, G.M. and Hillis, R.R. (2002). Assessing a basin's potential for geological sequestration of carbon dioxide: An example from the Mesozoic of the Petrel Sub-basin, NW Australia. In *The Sedimentary Basins of Western Australia* 3: 439–463. www.netl.doe.gov/sites/default/files/2018-10/ATLAS-V-2015.pdf.

Herzog, H.J. (2018). *Carbon Capture*. Cambridge, MA: MIT Press.

IEA (2013). Methods to assess geologic CO_2 storage capacity: Status and best practice. www.iea.org/reports/methods-to-assess-geological-co2-storage-capacity-status-and-best-practice.

IPCC (2005). Underground geological storage. In B. Metz, O. Davidson, H.C. de Coninck, M. Loos and L.A. Meyer (eds), *Special Report on Carbon Dioxide Capture and Storage*, 196–276. Prepared by Working Group III of the Intergovernmental Panel on Climate Change. New York: Cambridge University Press.

IPCC (2018). IUPAC-NIST Solubility Database, NIST Standard Reference Database 106. Geneva: World Meteorological Organization. Available from https://srdata.nist.gov/solubility/sol_detail.aspx?goBack=Y&sysID=62_172.

Kelemen, P. et al. (2019). An overview of the status and challenges of CO_2 storage in minerals and geological formations. *Frontiers in Climate*, November 15. Available from https://doi.org/10.3389/fclim.2019.00009.

Kjolhamar, B. et al. (2021). New approaches to CCS. *First Break* 39 (6): 53–57. https://doi.org/10.3997/1365-2397.fb2021043.

Lysyy, M., Ersland, G. and Fernø, M. (2022). Pore-scale dynamics for underground porous media hydrogen storage. *Advances in Water Resources* 163 (2022): article 104167. https://doi.org/10.1016/j.advwatres.2022.104167.

Mouli-Castillo, J. et al. (2021). Enabling large-scale offshore wind with underground hydrogen storage. *First Break* 39 (6): 59–62. https://doi.org/10.3997/1365-2397.fb2021044.

National Academies of Sciences Engineering Medicine (2019). *Negative Emissions Technologies and Reliable Sequestration: A Research Agenda*. Washington, DC: National Academies Press. Available from www.nap.edu/catalog/25259/negative-emissions-technologies-and-reliable-sequestration-a-research-agenda.

Qian, Y., Wattenbarger, R.A. and Scott, S.L. (2004). Aquifer gas injection offers an alternative for handling associated gas produced from deepwater fields. Paper prepared at the Offshore Technology Conference in Houston, Texas, May 3–6, OTC 16254.

Scottish Carbon Capture and Storage Group (2015). *Optimising CO_2 Storage in Geological Formations: A Case Study Offshore Scotland. CO_2MultiStore project*. Edinburgh: University of Edinburgh. Available from www.sccs.org.uk/images/expertise/reports/co2multistore/SCCS-CO2-MULTISTORE-Report.pdf.

Spivey, J.P., McCain, W.D. and North, R. (2004). Estimating density, formation volume factor, compressibility, methane solubility, and viscosity for oilfield brines at temperatures from 0 to 275 °C, pressures to 200 MPa, and salinities to 5.7 mole/kg. *JCPT* 43 (7): PETSOC-04-07-05. https://doi.org/10.2118/04-07-05.

Spycher, N. and Pruess, K. CO_2-H_2O mixtures in the geological sequestration of CO_2. II. Partitioning in chloride brines at 12–100°C and up to 600 bar. *Geochimica et Cosmochimica Acta* 69 (13): 3309–3320. https://doi.org/10.1016/j.gca.2005.01.015.

Steel, L., Liu, Q., Mackay, E. and Maroto-Valer, M.M. (2016). CO_2 solubility measurements in brine under reservoir conditions: A comparison of experimental and geochemical modeling methods. *Greenhouse Gases: Science and Technology* 6 (2): 197–217. https://doi.org/10.1002/ghg.1590.

Thomas, C., Loodts, V., Rongy, L. and De Wit, A. (2016). Convective dissolution of CO_2 in reactive alkaline solutions: Active role of spectator ions. *International Journal of Greenhouse Gas Control* 3:230–242. https://doi.org/10.1016/j.ijggc.2016.07.034.

US Energy Information Administration (2022). Underground natural gas storage capacity (webpage). www.eia.gov/dnav/ng/ng_stor_cap_a_EPG0_SAC_Mmcf_a.htm.

White, J. et al. (2018). *4D Seismic – Monitoring and Modelling*. British Geological Survey, Natural Environment Research Council. Available from www.sintef.no/globalassets/project/pre-act/documents/public-wp3-workshop-utrecht/2018.02.21_bgs-white_4d-seismic-monitoring-and-modelling.pdf.

Williams, G. and Chadwick, R.A. (2017). An improved history-match for layer spreading within the Sleipner plume including thermal propagation effects. *Energy Procedia* 114 (2017): 2856–2870. Available from www.researchgate.net/publication/319196019_An_Improved_History-match_for_Layer_Spreading_within_the_Sleipner_Plume_Including_Thermal_Propagation_Effects.

Wu, Y. and Li, P. (2020). The potential of coupled carbon storage and geothermal extraction in a CO_2-enhanced geothermal system: A review. *Geothermal Energy* 8: article no. 19. Available from https://doi.org/10.1186/s40517-020-00173-w.

Yekta, A.E., Manceau, J.C., Gaboreau, S., Pichavant, M. and Audigane, P. (2018). Determination of hydrogen-water relative permeability and capillary pressure in sandstone: Application to underground hydrogen injection in sedimentary formations. *Transp. Porous Media* 122 (2): 333–356.

Zhang, J., Li, Y. and Yu, H. (2019). Review on technologies of compressed air energy storage in aquifers. IEEE 3rd Conference on Energy Internet and Energy System Integration (EI2), 1053–1057.

Zheng, J. and Mao, S. (2019). A thermodynamic model for the solubility of N2, O2 and Ar in pure water and aqueous electrolyte solutions and its applications. *Applied Geochemistry* 107 (2019): 58–79. https://doi.org/10.1016/j.apgeochem.2019.05.012.

Postface

As discussed in the introduction to this work, our aim in publishing this book is to fill a gap in the petroleum literature on aquifer characterization and modeling. In approaching this task, we have sought to deploy a multidisciplinary approach underpinned by timely, real-world knowledge and experience about the subsurface. Throughout this book, we followed a straightforward blueprint that weaves together and distills insights derived from geological, geophysical, petrophysical and reservoir engineering disciplines, striving to show how the existing knowledge on reservoir characterization and modeling can be applied to the task of characterizing most aquifers.

Aside from the utility such knowledge holds for more accurate oil and gas production forecasts, it can also inform the current interest in the potential of aquifers to serve as storing sites for gases such as CO_2, CH_4, H_2, air, etc. The tools that we introduced in Chapter 9 should enable readers to arrive at a better, more informed take on the horizon of possibilities such aquifers hold for gas storage. It has become clear that companies looking to play an active role in the green energy (CCS, H_2, CAESA, geothermal, etc.) transition will need the technical expertise of people equipped with classical petroleum geoscience and engineering skills.

In Chapter 1, we introduced some misconceptions that concern aquifers. Now we can have another look at them through the lens of lessons learned in other chapters:

- When it comes to aquifers, both size and connectivity are important.
- Aquifers can reliably denote increased HC reserves only when good transmissibility is present between them.
- The FWL is the same as the OWC only in cases where the transition zone is minimal.
- Aquifer properties (salinity, porosity, permeability, etc.) are functions of depth and generally degrade as aquifers become deeper, except for pressure.
- Paleo zones in aquifers can contain residual hydrocarbon.
- Hydrostatic gradient in an aquifer is approximately 0.433 psi/ft for pure water only. The impact of salinity and depth needs to be taken into account when estimating this gradient.
- OWC can be tilted when in the presence of dynamic aquifers.
- Oil always occurs above the water leg, except when there are perched water zones.
- Aquifer permeability can be estimated accurately through logs unless there are heterogeneities such as fractures in the container rock.

- Drilling an injector into an aquifer is better than drilling it into an oil leg. This depends on the permeability change versus depth and the relative permeability to water in the oil leg.
- The optimal gridding of an aquifer is a function of the size of the aquifer and its transmissibility.
- An aquifer can be "outrun" in a gas reservoir, especially when the gas production rate exceeds the aquifer influx rate.

It is our view that the workflows and guidelines presented in this book are quite robust and can be applied in almost all cases where an aquifer needs to be characterized.

PARTING THOUGHTS

The greatest challenge facing mankind today is the transition to a more sustainable energy infrastructure and the concurrent mitigation of greenhouse gas emissions. Meeting this challenge will require a diversified array of solutions spanning multiple industries. One of the solutions rising to the fore is the potential to rapidly build out carbon sequestration, which involves the removal of CO_2 from the atmosphere and its storage in the subsurface. Carbon capture and storage has the benefit of being able to directly build on the extensive physical, capital and human infrastructure of the oil and gas industry. We have drawn on in-depth knowledge of subsurface fluid flow, and provided a modern interdisciplinary perspective, to arrive at a practical technical guide into the potential that aquifers hold as sites for carbon storage.

Aquifers occupy a significant part of the Earth's available volume in the subsurface and thus hold immense potential as sites for carbon storage. Many aquifers have been studied extensively as part of oil and gas energy development projects and, as such, they represent an opportunity to sequester carbon within existing areas of infrastructure that have already been impacted by, and integrated into, an inherited energy framework. Moreover, future efforts to reconfigure the landscape of our national and global energy systems can extract valuable lessons from this existing trove of data and expertise.

Appendix A: Units: definitions, prefixes and conversions

1 Da = 0.987 10^{-12} m²
1 lb/ft³ = 0.062428 kg/m³
1 poise = 0.1 Pa.s
1 psi = 0.0001450 Pa

bar	measure of pressure equal to 0.987 atmosphere or 14.5 psi
bbl	barrel
Bcf	billion standard cubic feet
bn	billion
bn bbl	billion barrels
bw	barrel of water
CO_2 density	1562 kg per m³ equals 44.25 kg per ft³
degree API	a measure of oil density, where 'API' refers to API gravity given by the following formula: API gravity = (141.5/specific gravity) - 131.5
Etpa	exaton per annum equal to 10^{18} tons per year
ft	foot
ft³	cubic foot
Gt	Gigaton, 10^{12} kilograms, 10^9 tons, 10^3 Megatons
$GtCO_2e$	Gigaton carbon dioxide equivalent M thousand
Gtpa	Gigaton *per annum*, 10^9 tons per year
ha	hectare
km	kilometer
kPa	kilopascal
k ppm	thousands of parts per million
Ktpa	kiloton *per annum*, 10^3 tons per year
m	meter
m³	cubic meter
m/s	meters per second
Mboe	1000 barrels of oil equivalent
MCF	1000 standard cubic feet
md	millidarcy
MM	million
MMboe	million barrels of oil equivalent
MMbw	million barrels of water

ms	millisecond
Mt	megaton, 10^6 tons
Mtpa	megatons per annum, 10^6 tons per year, or 61.9 MCF of CO_2 per day
MWe	Megawatt electric
MWh	Megawatt-hour, 10^6 watts per hour
Ω-m	ohm-meter
Pa	pascal
ppm	parts per million OR parts per meter
ppt	parts per thousand
psf	pound-force per square foot
psi	pound-force per square inch
p.u.	porosity unit
SCF	standard cubic feet
STB	stock tank barrel
T	a metric ton or 1000 kg OR trillion
TWh	Terawatt per hour, 10^{12} watts per hour
W	Watt
Wh	Watt-hour, 3,600 Joules
yr	year

Appendix B: Acronyms and abbreviations

2D	two-dimensional
3D	three-dimensional
4D	four-dimensional
AAPG	American Association of Petroleum Geologists
A/B	amplitude/background ratio
AI	acoustic impedance
AII	aquifer influx index
AMCOR	Atlantic Margin Coring Project
A_q	aquifer volume
AQ	Aquifer
AQ:HC	aquifer to hydrocarbon pore value ratio
atm	atmospheric pressure
AVA	amplitude-versus-angle
B_g	gas formation volume factor / gas expansion factor
CAESA	compressed air storage in aquifer
BHP	bottomhole pressure
BHPB	Broken Hill Proprietary Billiton Ltd.
B_o	oil formation volume factor / oil volume factor
CAESA	compressed air storage in aquifer
CCS	carbon capture and sequestration
CI	colored inversion
CO2-SCREEN	CO2 Storage prospeCtive Resource Estimation Excel aNalysis
CSEM	controlled source electromagnetic
DAS	distributed acoustic sensing
DBML	depth below mudline
DHR	depleted hydrocarbon reservoirs
DoE	US Department of Energy
DRP	digital rock physics
DST	drill stem tester
DTS	distributed temperature sensors
EEI	extended elastic impedance
EI	elastic impedance
EIC	Energy Infrastructure Council
EM	electromagnetic
EOD	environment of deposition

EOR	enhanced oil recovery
E&P	exploration and production
EPA	US Environmental Protection Agency
E–W	east–west
FDP	field development program or plan
FI	fluid impedance
FID	field investment decision
FLAG	Fluid Application of Geophysics
FWI	full waveform inversion
FWL	free water level
GDE	gross depositional environment
GI	gradient impedance
GIIP or OGIP	gas initially in place / original gas in place
GOC	gas–oil contact
GoM	Gulf of Mexico
GOR	gas-to-oil ratio
GRV	gross rock volume
GWC	gas–water contact
HC	hydrocarbon
HCPV	hydrocarbon pore volume
HO	Havlena-Odeh
ICE	integrated container edge (maps)
IEA	International Energy Agency
IMPES	implicit pressure, explicit saturation
IODP	International Ocean Discovery Program
IOR	improved oil recovery
IPCC	Intergovernmental Panel on Climate Change
ISO	International Organization for Standardization
LI	lithologic impedance
LWD	logging-while-drilling
MBAL	Material Balance (program)
MDT	modular dynamic tester
MM or M	Million
MPD	measured pressure data
MT	magnetotelluric
NaCl	sodium chloride
NE	northeast
NETL	US Department of Energy's National Energy Technology Laboratory
NIST	National Institute of Standards and Technology
NMO	normal moveout correction
NMR	nuclear magnetic resonance
NNC	non-neighbor connection
NRC	US National Research Council
NREL	US National Renewable Energy Laboratory
N–S	north-to-south
NtG, NTG, N:G	net-to-gross

NW		northwest
OBN		ocean bottom node aquisition
O&G		oil and gas industry
OGIP		Original gas in place
OOIP		original oil in place
OWC or WOC		oil–water contact
OWIP		original water in place
P_c		capillary pressure
P_g		Probability of geological success
PDF		probability density function
PEM		petro-elastic model
PGOC		possible gas–oil contact
PNC		pulsed neutron capture
PRMS		Petroleum Resources Management System
PSDM		pre-stack data migration
PTA		pressure transient analysis
PV		Pore volume
PVT		pressure-volume-temperature OR production validation testing
P-wave		primary wave or pressure wave
QI		quantitative interpretation
RCA		routine core analysis
RF		recovery factor
RFT		repeat formation tester
RMS		root-mean-square
RPM		rock physics modeling
RPT		rock physics template
RQ		rock quality
RQC		Rock Quality Consortium for Quantitative Prediction of Sandstone Reservoir Quality
RSWC		rotary sidewall core
R_t		unflooded zone resistivity
R_{xo}		flooded zone resistivity
SA		saline aquifers
SAII		specific aquifer influx index
SCAL		special core analysis
SCCS		Scottish Carbon Capture and Storage research group
SCF		Standard cubic feet
SDWA		Safe Drinking Water Act
SE		southeast
SEG		Society of Exploration Geophysicists
SEM		shared Earth model
S_g		gas saturation
SGR		shale gouge ratio
SGS		sequential Gaussian simulation
SI		shear impedance
SME		subject matter expert

S/N	signal-to-noise ratio
SNA	sum of negative amplitudes
S_o	oil saturation
SPE	Society of Petroleum Engineers
SRC	seismic reservoir characterization
SRMS	storage resources management system
STB	Stock Tank Barrel
S_w	water saturation
SW	southwest
S-wave	shear wave
SWC	sidewall core
S_{wirr}	irreducible water saturation
TCF	trillion cubic feet
TDS	total dissolved solids
TVD	True vertical depth
TVT	true vertical thickness
UAE	United Arab Emirates
UHS	underground hydrogen storage
UIC	Underground Injection Control Program
USGS	United States Geological Survey
VEH	Van Everdingen and Hurst
VRR	voidage-replacement ratio
VSP	vertical seismic profile
W–E	west-to-east
WOC	water–oil contact
XRD	X-ray diffraction
XRF	X-ray fluorescence

Appendix C: Aquifer characterization and modeling checklist

The following list contains many items that should be considered when characterizing and modeling aquifers. They are not all relevant in all situations; however, the study team is encouraged to review this list to ensure that they have not overlooked any of these considerations.

Sources of aquifer characterization data

- High quality seismic imaging
- Well penetrations with high quality logging and with SCAL testing on core
- Static and dynamic pressure surveillance
- Reservoir fluid sampling and PVT testing
- Production history
- Reservoir and aquifer property geomodeling and Touchstone modeling
- Properties and performance of analogue reservoirs

Aquifer description

- Locate field within the basin/reservoir system (and other fields within same aquifer).
- Gross and net (permeable) reservoir thickness map(s)—seismic or well-based ideally defined by seismic mapped top and base or isopach and perhaps inversion-based sand distribution.
- Fault pattern and throw estimates through the aquifer i.e. reservoir juxtaposition and expected flow baffles/barriers.
- Well correlations within the aquifer and through the field to judge connectivity and continuity.
- Define potential aquifer segments (generally fault bounded) between which properties may vary and/or flow be confined.
- Pore volume map(s) of the aquifer (used to describe Size: HCPV to aquifer PV).
- Porosity vs. depth trends across the aquifer—well derived or diagenetic model derived ("Touchstone").
- Permeability vs. depth trends across the aquifer—well derived or diagenetic model derived ("Touchstone").
- Hydrodynamic, dynamic or overpressured reservoir?
- Pressure vs. depth relationships, pressure sinks, leak points, hydraulic head plots and maps, Darcy Flux (flow/unit area) estimates if hydrodynamic system.

- Aquifer salinity and other water chemistry data and regional variation along with expected temperature variation.
- Proximity of other potential aquifers into which injected water may leak over time or cross-flow with excessive injection/fracturing?
- Presence of transition zones or bitumen mats that could impede aquifer influx?

Aquifer model construction

- Map of aquifer extent.
- Identify communication barriers/baffles between aquifer extent and completed intervals in production wells.
- Map aquifer properties and their variations.
- Specify series of numerical aquifer tanks with average properties derived from property maps.
- Determine water–oil (or water–gas for gas reservoirs) relative permeability, particularly water mobility and displacement efficiency in the transition zone.
- Determine fluid saturations as functions of depth (including in the aquifer).
- Quantify any gas dissolved in the aquifer brine calibration and history-matching.

Numerical aquifer pitfalls to avoid

- A single large numerical aquifer cell can place large brine volumes unrealistically close to a reservoir boundary. (Instead, use a series of numerical aquifer cells arranged to reflect mapped aquifer volumes.)
- A single numerical aquifer can be connected along an extensive reservoir model boundary. This provides instantaneous communication over very large distances. (Instead, use multiple numerical aquifers, and check the connections to reservoir boundaries.)
- Petrel can generate erroneous connections of numerical aquifers to interior reservoir grid blocks. (Generation of aquifer connections with Petrel should be avoided or its results must be corrected by hand.)
- History-matching often results in a minimum aquifer size. This may also show up as a reservoir pressure fall-off at the end of production history. Long-term forecasts may be pessimistic as a result. (Use a range of aquifer sizes from the minimum history-matched size as a low case and to the mapped aquifer size as a high case).
- In the absence of well penetrations, too often aquifer rock properties are assumed to be identical to reservoir rock properties. (In some cases, aquifer properties can differ substantially from properties in the hydrocarbon column and can degrade significantly with depth. Be sure to use the best geoscience estimates of aquifer properties and their depth trends.)
- Any remaining hydrocarbon saturation in the aquifer pore space can significantly reduce aquifer strength due to relative permeability effects. (Don't assume the effective aquifer permeability matches reservoir permeability, and don't assume injectivity into the aquifer without first testing.)

Index

acoustic impedance 8, 58–62, 209
aggregation 39–40, 115
AII (aquifer influx index) 143, 146, 148–50, 209, 211
air, solubility in water 201
air, storage 165
air storage 199
Alberta 104, 166, 190–1
Algeria 166
amplitudes 15, 17–21, 27–8, 53, 55, 60, 144, 187
amplitude-versus-angle (AVA) 17, 209
analysis: geochemical 186; hydrodynamic 171; petrophysical 63, 100; quantitative 49
analytical modeling 123, 125, 127, 129
anisotropy 104, 170
anomalous water 79–80
AQ: HC 121, 136, 140, 150, 152–4, 159, 161, 163, 183, 209
aquifer: analytical modeling 121; analytical models 136; approach 136; area 136; boundary 130; brine 133; cases 164; cases for Field C 164; categories 150; cells 133; compressibility 152; connections 214; connectivity 140; definition 53; description 5; detection 3; drive 126; effectiveness 2; encroachment 70; extent 31; geometry 117; gradient 45; grid blocks 141; hydrogen storage 198; parameters 139; performance 121; pore volume estimating 163; pressures 29; properties 1; rock properties 140; salinity 25; size 2; strength 62; system 21; types 141; volume 2
aquifer boundary 134
aquifer cells 134–5
aquifer characterization 3, 37, 51–2, 148, 157, 165, 168, 170, 205, 213; workflow 146
aquifer connectivity 146, 148
aquifer description 37–87, 213
aquifer description, workflow 143
aquifer detection 21
aquifer drive 153
aquifer effectiveness 72, 126, 144–5, 148, 152

aquifer encroachment 139–40
aquifer extent 133, 214
aquifer geometry 121, 133, 141
aquifer impact 137, 150, 152
aquifer influx 117, 121, 136, 143–55, 190, 214; index 143; index categories 151; problem 121; rate 148; volume 150
aquifer influx index *see* AII
aquifer influx rate 150, 206
aquifer modeling 3, 138, 213
aquifer models 105, 112, 130, 138; analytical 133; conventional grid 136; numerical 138
aquifer parameters 164
aquifer performance 133, 141
aquifer pressures 48, 85, 130, 137, 145, 148
aquifer properties 114, 133, 140, 157, 205, 214
aquifer rock properties 214
aquifers: poor 149–50; regional 42, 80, 192; tilted 29; weak 153
aquifer salinity 177, 214
aquifer size 32, 72, 121, 123, 140, 143, 152–4, 214
aquifer strength 121, 125, 146, 153–4, 190–1, 214; index 121; quantification 148
aquifer strengths, typical GoM fields 151
aquifer system 25–6, 39–40, 50, 157–8, 200
aquifer types 150
aquifer volume 121, 123, 134, 136–7, 139, 146, 191–2, 209
Archie's law 23, 25
areal extent 27, 53, 143, 146, 159, 163
Atlantis field 57
Australia's northwestern shelf 52
AVA (amplitude-versus-angle) 17, 209

baffles 47, 50–1, 54, 56, 61, 112, 117, 134, 140, 144, 159, 161
baffling 117, 121, 133, 141
barriers 41, 47, 56, 112, 117, 134, 161
basalt formations 165
basin 27, 37–41, 43, 85, 93, 99–100
Basin margins 30–1

215

216　Index

Basin modeling 3, 93–4, 98–100, 102–3, 105, 157; workflows 94
Basin models 93–7, 99–101, 105
Bass Strait 42, 45
Blane field 86
boundaries 52, 64, 109, 134, 137, 143, 194
brine 13, 140, 173, 177, 181–2, 187, 191–2, 195
brine salinity 187
burial history 94–5, 102, 157

CAESA 165, 199–200, 205
calibration 10, 19, 57, 59, 94, 98, 104, 106, 169, 171
capacity, effective reservoir 190
capillary entry pressure 48, 84, 94, 173, 181
capillary pressure 49, 68, 84, 96, 182, 198, 211
capillary seals 94, 101
caprock 42, 167, 173, 181–2, 186, 188, 198–9
Carter-Tracy 129, 136–38
Carter-Tracy influence functions 130
CCS 3, 165–6, 205, 209
cementation 8, 21, 100, 102–4
CH_4 solubility 195–6
CH_4 solubility in brine 195
chemical reactions 188, 192, 198
classification, geologic storage formation 40, 42
class-VI disposal wells 167
CO_2 63, 165–73, 177, 179, 181–8, 190–4, 198, 205, 208; containment in storage 177; density 173; dissolution in brine 177; dissolved 177, 182–3; injection 165; phase change 193; plume 173; sequestration capacity 172; solubility in brine 177; storage 3; viscosity 173; volume change 176
CO_2 and water density 174
CO_2 and water viscosity 175
CO_2 density 189, 207
CO_2 injection 170, 181, 183–4, 186, 190, 194, 198
CO_2 injection for EOR 194
CO_2 injectivity 172, 194
CO_2 plume 177, 183–5, 194
CO_2-SCREEN 172, 209
CO_2 sequestration capacity 190–1
CO_2 solubility 177–9
CO_2 storage 165–7, 170, 172, 177, 183, 190, 195, 198; capacity 167; mechanisms 189
CO_2 storage capacity 189
CO_2 Storage prospeCtive Resource Estimation Excel aNalysis 209
CO_2 storage sites 165, 167–8, 182, 185
CO_2 surface leaks 183
CO_2 viscosity 176
cole plot 124, 126–7
communication 7, 114, 134, 139–40

compaction 8, 30, 93, 101
comparison of aquifer models capabilities 138
compartmentalization 8, 43, 48, 69, 117, 173
compartments 39, 48, 110, 171, 187
compressed air energy storage 199–200
compressibility 2, 8, 122–3, 133, 138, 177
connected aquifers 60, 140
connectivity 2, 8, 29, 32, 46, 48, 54, 140, 143–4, 146, 158–9, 161
Connolly's method 18
core analyses 49, 103, 106; special 11, 49, 105, 134, 171, 211

decatur project 166, 184
depleted hydrocarbon reservoirs (DHR) 165, 209
depleted reservoirs 189–90
depletion 2, 7, 42, 45, 70, 121, 124, 126, 130, 148–51, 190
depth 1, 6–7, 10–11, 26, 69–70, 74, 79–80, 101–4, 109, 153–4, 157–9, 163, 173, 176–7, 205–6, 214
DHR (depleted hydrocarbon reservoirs) 165, 209
diagenesis 8, 21, 50–1, 93, 99–104, 106, 158; ancient 103; basic 100; clay 8; mineral 31; significant 18; simulating 99; thermal 157
digital rock physics (DRP) 105, 209
dissolution 102–4, 183–4, 192–3, 198
distributed temperature sensors (DTS) 185, 187, 209
drilling 1, 9–10, 194, 200, 206
DST (drill stem testers) 11, 74, 209
dynamic data 30, 54, 117
dynamic modeling 140, 157–8, 163
dynamic models 93, 112, 114, 133, 163, 170

East Mediterranean 10
ecosystems 183, 187
EEI (extended elastic impedance) 17–19, 209
effective stress 21, 99, 101–2, 157
Ekofisk 87
elastic impedance (EI) 17–18, 209
elastic properties 8, 57–58, 144
El Bagoury 66
electromagnetic, controlled source 177, 209
encroachment volume 139–40
entry pressure 68–9, 71, 182
environment 52, 103, 184, 209
EOR (enhanced oil recovery) 9, 166, 169, 194, 209
EOR applications 166, 170
equilibrium, hydrodynamic 26, 84
estimate rock properties 105
estimating aquifer contribution 124, 126
EUR (estimated ultimate recovery) 150
evolution of fluid pressures 62, 96

facies 17, 48–51, 58–60, 82, 102, 106, 109–10, 144, 146
facies classification 50, 56, 59, 106
facies inversion example 59
faulting 8, 43, 47, 52–3, 81, 115, 136–7, 143, 145
fault interpretations 108–9
fault juxtaposition 54, 117
faults 26, 28, 47–8, 50, 53–4, 61, 87, 107, 109, 112, 114, 117, 134, 170–3, 187
fault seal analysis 47
fault transmissivity 54, 112, 117, 144
FDP (field development plan) 27, 60, 112, 115, 210
Fetkovich 138
Fetkovich aquifer 130, 137
FI (fluid impedance) 8, 210
field case studies 141, 157, 159, 161, 163
field example of perched water 81
flow 6–7, 42, 47–8, 65, 67, 69, 85, 87, 101, 104, 161, 196, 198
flow barriers 50, 172
flow units 50, 67, 112, 144
fluid contacts 28, 55–6, 110, 112, 143
fluid flow 42, 48, 51, 93, 96, 100–1, 112, 144
fluid movement 47, 56, 60–1, 93, 100, 102, 157
fluid pressures 60, 62, 94, 96, 99, 101, 146
fluid properties 8, 21, 51, 56, 65, 105, 144, 172, 192
fluid property analysis 106
fluids 8, 10, 20–1, 29, 31, 47–8, 58, 60–1, 65, 67–8, 93–4, 98–9, 104–6, 117, 181; residual 106
fluid sampling 10, 14, 55, 64
fluid saturation determination 106
formation 11, 14, 31, 65, 68, 166, 171, 181, 184, 188, 194
formation testers 64, 211
formation water 31, 74; resistivity 13
formation water resistivity 25
frac pressure 199
fractures 42–3, 65, 104, 170, 173, 181, 183, 187, 205
frequency 15–16, 18, 20–1, 27–8, 60, 144, 184
fresh water 6, 11–12, 26, 177
Froan Basin area 193
full waveform inversion (FWI) 16, 55, 210
FWI (full waveform inversion) 16, 55, 210
FWI-derived seismic velocities 57
FWI technologies 16
FWL (free water level) 1–2, 68–9, 71, 73–4, 76–7, 121, 139, 205, 210
FWL and WOC 69

gas saturation, trapped 70–3, 181, 192
gas storage 165, 197, 199, 205
GDEs (gross depositional environment) 42, 45, 210

geological facies 48, 67, 110, 170
geomechanics 100, 103
geomodeling 105, 112; aquifer property 134, 213
geomodeling workflow 106
geomodels 105–6, 110–13, 115, 144
geophysical input to aquifer size 143
geophysics 1, 5, 37, 57, 103, 184
GoM (Gulf of Mexico) 38–9, 46, 143, 150, 157, 192, 210
GoM fields 62, 140, 143
gravity 14, 67, 177, 183, 207
grid cells 133–4, 136
gridding 107, 134
groundwater 6–7, 21, 80, 85
GRV (gross rock volume) 53, 143–5, 163
GWC 55, 68–9, 79, 110

Havlena-Odeh method 123, 125, 210
heterogeneity 49, 51, 68, 104, 112, 117, 146, 190, 192, 205
history-matching 123–4, 133, 140, 150, 214
hydrogen 63, 166, 198–9

ICE (integrated container edge) 53–4, 171, 210
ICE maps 53, 143
imaging 64–5, 188
imbibition 177, 196
infinite aquifers 2, 70
injection 63, 151, 166, 171–2, 177, 181, 183–4, 189, 194, 200
injectivity 153, 168, 170, 181, 214
integration 29, 51, 57, 100, 105, 158
Iraq 31
isolated compartments 79–83

key aquifer parameters 150
key CCS projects 166

lithology 18, 21, 41–2, 58, 64–5, 170, 172
Lorentz plots 67–8

mapping 5, 7, 10, 15, 28, 107, 109, 112, 133
material balance 130–1, 137
material balance equations 124, 127
maximum aquifer volume 133
maximum injection pressures 194
MBAL model 138–9
MBAL water influx modeling for history-matching 124
MDT (modular dynamic testers) 11, 31, 68, 94
modeling checklist 3, 213
models: analytical 137–39, 148; diagenetic 101, 213; numerical 3, 67, 138; pot-aquifer 148; unsteady-state 148
monitoring CO_2 storage 186

NahrUmr Lower Sands accumulation 31
natural-gas-storage 169, 195, 197
net rock porous volume 145
NMR (nuclear magnetic resonance) 10, 65, 77
NtG (net-to-gross) 3, 21, 67, 110, 115, 144–5, 148, 158, 163, 210
numerical aquifer cells 134–5, 139–40, 158, 214
numerical aquifers 133–4, 139–40, 163, 214; connected 139; multiple 135–6, 140, 214

Ogallala aquifer 21, 26
oil field offshore Angola 27
overburden 20, 56–7, 94, 99, 170, 172, 181
overburden properties 9, 170
overlying Miocene Utsira Formation 185
OWC 1, 28–30, 53, 55–6, 68, 71, 79–80, 82–3, 143–4, 205, 210

paleo zone 64, 70, 76
PEM (petro-elastic model) 57, 211
perched water 79–82, 85
perched water identification 82
permeability 1–3, 48, 51, 65–7, 85, 87, 101, 104–5, 110, 153–4, 157–9, 161, 163, 170–3, 199
permeability reduction 153, 198
permeability trend 157, 159
petroleum reservoir fluids 14
phase behavior 14
plume 177, 181, 183, 193
pore volume (PV) 70, 72, 121, 139, 163, 188, 210, 213
porosity 21, 25–6, 49–52, 64–5, 67, 73, 75–6, 101–4, 110, 114–15, 144–6, 157–9, 161, 163, 171–3, 198–9; determination 64; predictions 102, 103; preservation 102; trend 158, 163
porosity-permeability trends 50, 67
pressure 2, 9–10, 53–4, 68, 94, 96, 99–101, 121–2, 138–40, 144, 166–7, 171–5, 177–8, 189–90, 194–5, 198–9, 201–2, 205; bottomhole 140, 209
pressure and temperature 10, 94, 100, 144, 198
pressure buildup 140, 159
pressure data, measured 31, 210
pressure depletion 117, 122, 137
pressure gradient 42, 84–6; typical aquifer 85
pressure response 130, 137
pressure sources 94, 96
pre-stack data 17, 20, 28
PRMS (Petroleum Resources Management System) 117, 169, 211
production technology 66
productivity index, steady-state 130, 137
PVT (pressure-volume-temperature) 8, 211, 213
P-waves 15–17, 177, 211

QI (quantitative interpretation) 17, 20, 61, 211

rate of pressure buildup 140
RCA (routine core analysis) 105, 211
relative permeability 70, 72–5, 77, 96, 99, 106, 112, 177, 181, 196–9, 206
reservoir: boundaries 123; cells 133; connectivity 117; facies 56
reservoir & aquifer characterization workflow 146
reservoir boundaries 133, 135–6, 138, 140, 192, 214
reservoir cells 136–7
reservoir characterization 1, 64, 169, 171, 205; seismic-based 9
reservoir complexity 65, 173
reservoir conditions 77, 121, 137, 148, 177, 189, 192–3
reservoir connectivity 140
reservoir facies 57, 59
reservoir fluid sampling 134, 213
reservoir heterogeneity 68
reservoir monitoring 65
reservoir performance 134
reservoir pressure 2, 45, 140, 148, 150, 190, 214
reservoir properties 8, 105, 173
reservoir quality (RQC) 48–9, 65, 93, 99–102, 114, 211
reservoir quality, models 93, 100
reservoir quality, predictions 99, 101–2
reservoir rock properties 140, 214
reservoir rocks 8, 50, 100, 181, 183, 192, 194
reservoirs 1–3, 15, 40–2, 47–51, 53–7, 61–2, 67, 69–71, 77, 79, 93–122, 129–31, 133–7, 140–4, 148–52, 165–7, 169–73, 181, 192–4; analogue 134, 213; petroleum 68, 121–2, 181; subsurface CO_2 storage 168
reservoir simulation 62–3, 106, 112, 117, 144, 146, 153, 159
reservoir simulators 136–37, 159; numerical 130, 192
reservoir systems 2–3, 5, 28, 157
resistivity 10, 13–15, 23, 25–6, 78, 110; electrical 14, 25, 177
resources 27, 38, 114–15, 117, 169
RF (recovery factor) 70, 115, 153–4, 189–91, 211
rock physics 100, 102–3
rock physics model (RPM) 55, 60, 211
rock properties 3, 21, 52, 69, 101–2, 106, 110, 144, 157
rock type 8, 50, 110, 197
RPT (rock physics templates) 57–9, 102, 144, 211
RQ (rock quality) 50, 100–1, 157, 211
RQC see reservoir quality

saline aquifers 3, 165–9, 172, 177, 182, 192, 196, 211
saline aquifers in North America 168

Index

salinity 12–14, 21, 23–6, 32, 171, 173, 177–8, 187, 192, 195–6, 198, 201, 205
sands 6, 24, 28–9, 46, 52, 54, 58–9, 67, 157, 163
sandstone reservoir quality 101, 211
SCAL 11, 49, 105–6, 134, 171, 211, 213
scales 5, 7, 26, 31, 37, 40, 43, 49, 56, 93, 110, 112, 143–4
Schilthuis model 148
seal capacity 94, 100–1, 171
seal permeability 171, 181
seals 7, 11, 42, 47–8, 51, 54, 87, 100, 103, 169–71, 173, 181–2, 186
section analysis, thin 106
sedimentary basins 30, 38, 100, 167
sedimentology 103, 106
sediments, clastic 6, 23, 101, 104
segments 20–1, 37–9, 54, 63, 115, 143–4, 158
segment scale 56, 60, 144
seismic 8, 14–15, 17, 19–21, 24, 27–9, 52–6, 58, 60–3, 107–10, 143–4, 185–7, 213; amplitudes 16, 20, 52–4, 185, 187–8; attributes 16, 20, 28, 55, 60
seismic data 17, 20, 24, 32, 70, 99, 109, 171–2, 177, 185–6; pre-stack 28–9; type 16, 54, 186
seismic example 27, 52
seismic interpretation 5, 15, 21, 51, 54, 103, 112
seismic inversion 19, 56–8, 99, 144
seismic reflections 16–17, 56
seismic reservoir characterization (SRC) 56–8, 60, 211
sequestration 3, 165, 167, 195, 209; capacity 188; capacity estimation 186; lifespan 172
SGR (shale gouge ratio) 48, 211
shales 28–9, 40, 42, 45, 58–9, 65, 69, 103, 200
Sidewall cores 49, 211
simulations 62, 86–7, 99, 109, 117, 136, 183–4; dynamic 159; sequential Gaussian 158, 211; stochastic 163
simulators 193; numerical 198
site characterization 40, 169, 171
sites for CO_2 storage 172, 177
in situ stresses 117, 181, 184, 200
Sleipner field 166, 185–6; project site 185
Snohvit project 166
solubility 177
solution plot 127, 129
sonic velocities 177, 179, 187
spacing 150, 170
SRC see seismic reservoir characterization
SRMS (Storage Resources Management System) 169, 211
static models 93, 112, 114, 117, 144, 158–9
static models for reservoirs and aquifers 97–117
storage 6–7, 145, 157, 167–8, 177, 185, 188, 191–2, 195, 206; carbon dioxide 169; geologic 40, 42

storage capacity 172, 181, 188, 190, 192–3
storage capacity estimation 186, 192
storage efficiency 192, 197
storage resource estimates 167
storage site requirements 169
storage sites 165, 170, 172, 180, 184, 186; potential 172–3, 181; stratigraphic complexities 52, 117, 143
stratigraphy 25, 27, 40, 42, 52, 170
strength 29, 32, 117, 121, 123, 133, 141–2, 146, 150, 152, 190–1
strength and effect of underlying aquifers 190–1
stress: monitor state of 64–5, 188; state of 51, 183
strong aquifer 149–50, 152–3
structural features 161, 170
surveillance 60, 65, 77, 185
system 2, 6–7, 24–6, 31, 41–2, 96, 98, 183, 200

tar mats 64, 77–9, 146
temperature and pressure 94, 101, 174–5
thin sections 49, 104
tilted OWCs 29, 86–7
top seal 24, 51, 53, 96, 101, 177, 181–2, 186; aquifer's 182
touchstone modeling 3, 134, 157, 163, 213
tracking CO_2 movement 187–8
transition zone 64, 68–71, 73–4, 77, 79, 83–4, 144, 146, 153, 205, 214
transition zone analysis 70, 105
transmissibility 117, 121, 133, 141, 152, 205–6
trap 5, 28, 39, 53, 83, 87, 107, 143, 171
trapped gas 70, 197
trapping mechanisms 169, 198

UHS (underground hydrogen storage) 165, 198, 212
uncertainties 3, 5, 8, 32, 58, 60, 115, 117, 146, 152
uncertainty analysis 31, 116, 163
underground hydrogen storage see UHS
underground storage 165, 195
underlying aquifers 190–1

van Everdingen and Hurst 122–3, 138
VEH (van Everdingen-Hurst) 123–4, 130, 137, 148
velocity 15–16, 21, 27–8, 60, 94, 96, 100, 144, 172, 177
voidage-replacement ratio 148, 212
volumes 17, 29, 39–40, 53, 58–9, 69–70, 115–16, 121, 124, 133, 141, 143, 188–9, 192–3; in-place 115
VRR 146, 148, 150–1

water: compressibility 70, 177; density 174; interstitial 2, 31; mobility 77, 133, 214;

original 148, 211; pure 177, 205; saline 13, 26, 177, 199
water contacts 1, 28, 53, 55, 70, 77, 83, 85, 183, 210
water encroachment 60–1, 122
water flow 6–7, 29–31, 63, 171
water influx 70, 122–4, 152
water injection 2–3, 62–3, 79, 85, 143–55; projects 143
water injection projects 152
water leg 1, 53, 105, 205
water movement 31, 61, 63, 139, 144
water production 68–9, 121
water resources 173
water salinity 9, 23–5, 65, 112, 171
water saturation 8, 25, 53, 58, 67, 76, 145, 190, 196, 211; irreducible 68, 104, 114, 212
water volume 80, 139–40, 146
Williston basin 102
WOC 31, 69, 73, 153–4, 210
workflows 3, 5, 37, 40, 43, 48, 98, 104–5, 110, 157–8, 161, 206

9781032224954